DO585216

COLLECTABLE TECHNOLOGY

THIS IS A CARLTON BOOK

Published in 2005 by Carlton Books Limited
An imprint of the Carlton Publishing Group
20 Mortimer Street
London W1T 3JW

A catalogue record for this book is available
from the British Library.

ISBN 1 84442 993 8

Executive Editor: Stella Caldwell
Editor: Peter McSean
Art Director: Clare Baggaley
Design: Mark Lloyd and Sooky Choi
Picture Research: Steve Behan
Production: Lisa Moore

Printed and bound in Dubai

CARLTON
BOOKS

WORKSTATION
PAGE 10

HOME BASE
PAGE 60

PLAYTIME
PAGE 108

INTRODUCTION

I remember life without mobile phones and personal computers, but a whole generation is too young to know what it was like. At school we used logarithm tables and drew graphs on lined paper; today's children use calculators and laser printers for the same tasks. The ability to contact anyone at any time by telephone or e-mail is taken for granted now, but in the not-so-distant past, a telephone call was a special event and the only mail system was the local postal service. The world was about to change. A utopian future was promised, where technology would be our servant and how to occupy ourselves in our free time would become a pressing concern.

It was in the 1970s that electronic technology gained the foothold that allowed it to climb into our daily consciousness. Back then, it manifested itself as digital watches – the ones with the red crystals – and the not-so pocket-sized calculators, with red LED displays. Not long after the arrival of pocket calculators came the debut of home computers, albeit in a hobbyist form at first. The home computer quickly developed into a more useful tool than its first calling as a games machine, killing off the humble typewriter and then coming to pervade all aspects of office life.

Over the past 40 years, the use of electronic technology in every aspect of our daily existence has risen to the point where many of us can't imagine life without it. Technology manages our time, helps us to communicate and even entertains us. Technological developments move fast and, as they mature, products converge. For example, mobile phones have built-in calculators, cameras, MP3 players and even Internet access. The mobile phone is now so important to modern life that telecoms visionary Hans Snook recently proclaimed that it would become "a remote control for life". Back in the home, the PC is becoming an entertainment centre capable of playing the role of television, video recorder and even DVD player.

1969
PHILIPS GF303 RECORD
PLAYER

Why do people collect technology? I believe there are three main reasons. Firstly, many of us are nostalgic, and as we get older, we pine after the items that remind us of our younger years. We want that old 1970s lava lamp which glowed in the corner of our bedroom, or that first Sony Walkman we listened to when we were supposed to be paying attention in school. The second reason is that humans are hoarders by nature. In the past there was a preoccupation with collecting stamps or coins, but nowadays, the thirty- and forty-somethings want a more sophisticated kind of object to collect. The final reason concerns the Internet, which in the past 10 years has come to permeate almost every aspect of our lives. Nearly everyone in the developed world has access to e-mail and the World Wide Web, and practically everyone is aware of auction websites. The proliferation of Internet auction sites has driven the recent market for desirable technology and, in many ways, was the inspiration for this book, along with my own fascination with electronic gadgets from the past and present.

What sorts of items are worth looking for and how do you spot them? I have come across collectors who are interested in almost every electronic item you can imagine. They range from full-sized commercial computers that can fill a room to the smallest radio capable of fitting inside your ear. Pocket calculators and electronic watches are particularly in demand,

as are items linked to specific inventors, such as Sir Clive Sinclair, the man famous for his pioneering work on computers and miniaturized televisions. Another group of collectors may be interested in the design of products rather than the technology, and items can be of interest to both this type of collector and the "gadget" collector. A good example of this is the Sinclair Executive pocket calculator, which was an electronic marvel as well as a design icon. When you are out and about seeking to buy items, a good rule of thumb is to look for the unusual. If you think you have an unusual find, something you don't often see, then the odds are that it's worth collecting and has some value.

1972
PANASONIC TOOT-A-LOOP

You can buy this desirable technology in a plethora of places, including garage sales, car boot sales, flea markets and junk shops. Many items change hands on Internet auction sites and this type of trading is becoming increasingly important. Of course, we shouldn't ignore the traditional auction rooms. Although auction houses are more used to dealing with antiques in the true sense of the word, they now increasingly handle contemporary objects. However, the sort of technology items in an auction house will be limited to those associated with a particular and famous designer or, for example, an item that can be linked to a famous event or person. It's worth noting that the value of an item, wherever you find it, is heavily dependent on condition. The items that command the highest prices are in as-new condition and come complete with the original packaging and instructions. Some items are so rare that non-functioning examples are in demand, although these are few and far between.

If you are new to collecting technology, there are some basic rules to follow. The first is to decide which area interests you most. If you don't, you will soon find yourself confused and overwhelmed. For example, you could start by collecting all the models of a particular type of electronic game or concentrate on all the variations or colours of a specific product type. When buying, always go for working examples in the best condition that you can afford – it will pay off when you come to sell your pieces. It is a good idea to take a set of batteries with you so that you can test equipment before buying it. Sellers will often say things work, even if they know they

1984
APPLE MACINTOSH CLASSIC

MOTOROLA 8500X CELL PHONE

don't! If you are buying off Internet auction sites then take note of the seller's rating or feedback. If other buyers have had cause to complain, then it's wise for you to shop elsewhere.

Many collectors actually make use of the items in their collections. For example, there are many enthusiasts of vintage audio equipment who use the systems they acquire and many collectors of vintage computers and games consoles who enjoy playing what they consider to be better games than the modern offerings available. This is one of the most enjoyable aspects of collecting technology – owning historic equipment that is not only good to look at, but actually does something useful.

I have divided this book into three sections based on Work, Home and Play, the main areas where technology has had a large impact on our lives. The world of desirable technology

can be considered limitless, and I have needed to be selective over which of the millions of available electronic items to include. I have tried to provide as broad a sample of the possibilities as I can, to illustrate the breadth of what you can find out there, and also to encourage you to experiment in your collecting.

PEPE TOZZO 2005

2000
BLUEROOM MINIPOD SPEAKERS

WORKSTATION

Advances in technology invariably enter the workplace first, having arisen as a result of industrial or military research. The NASA space programme, for example, is claimed to have pioneered many of the technologies that later found their way into everyday items – or so the advertising executives would have us believe. Many tools we use today – computers, fax machines and even cell phones, for instance – were initially the preserve of the workplace.

The first hints of technology change began in the 1960s, albeit in the wealthiest companies only. The price of big computers began to fall, bringing large-scale computing within the reach of smaller organizations. In the same decade, the first electronic calculating machines started to appear, but the relatively high price confined them to the wealthy few. The media latched on to the new electronic age and ran articles that predicted a utopian future, where everyone would have an excess of leisure time and robots would be our servants.

In the 1970s, the average office would still have relied on paper and people for most jobs. But as that decade progressed, more technology appeared in the workplace, mainly in the form of electronic calculators and, for companies that could afford it, the first simple computers. The original small computers were generally used for word processing tasks, instantly killing the typewriter.

The 1980s was the decade in which computing changed our lives forever. Office-based networking allowed workers to share information and to communicate using e-mail. The computer and printer became a more familiar sight. The arrival of portable computing hinted at further change, but the expensive and large luggable machines were really just executive playthings. Perhaps the biggest technological arrival was the cellular phone. Again just an executive plaything because of the huge cost, the new portable telephone system would change the way we communicated in the coming years.

The 1990s brought the Internet and World Wide Web to the public. Information became easily available to anyone with a personal computer and a modem, and e-mail began to overtake the humble paper letter. The way we shopped changed with the arrival of on-line stores. Cellular phones, now very affordable, were the gadget to have and rapidly became a lifestyle accessory. There was tremendous change in portable computing, firstly with the arrival of true laptop PCs, but then with the birth of the PDA (Personal Digital Assistant). It didn't take long for the manufacturers to start building PDAs into cellular phones, initiating a trend for product convergence that would put hugely powerful machines in our pockets during the next millennium.

Now, in the twenty-first century, we expect technology to do so much for us in our working lives. Technology has made the world a faster and smaller place. Communication takes seconds, computers handle telephone systems and we are always contactable on our cellular phones. There is nowhere to hide. That 1960s prediction of a utopian future, where our biggest problem would be how to fill our leisure time, seems further away than ever.

With such rapid transition in the workplace, the amount and rate of technology change provides collectors with a rich and varied selection of items. Practically everything is collected: computers, telephones, calculators, printers, cell phones and PDAs, to name just a few. The range of items available is difficult to comprehend. For example, the number of electronic calculator manufacturers in the 1970s would have been in the thousands. Most of these companies are long gone, but their products can still be found and make very interesting items for collecting.

Most work-related technology will never be as exciting as items from the home or play sectors, but with many examples still being useful working machines, it can be a rewarding area of exploration for the dedicated collector.

£300–£700 / US$560–$1,300

1964
FRIDEN EC-130

Many experts have described the Friden EC-130 as the world's second electronic calculator (the first was the 1961 ANITA Mark VII made in England). Friden, an American company, launched the EC-130 in 1964 at the staggering price of US$2,100, the equivalent of about $6,500 (£3,500) in today's money. The Friden was the first electronic calculator to use solid-state design, based on discrete components, and to display results on a CRT (cathode ray tube) screen. This was a revolutionary concept at the time, since the only other display option was to use Nixie tubes. The CRT screen allowed the display to show up to 13 digits, which was more than any rival machine could offer. Back then, the Friden was the most advanced calculating machine in the world, all other solutions being electro-mechanical. The basic functionality offered by the Friden can now be bought at your local store, for a small sum, in the form of an ordinary pocket calculator.

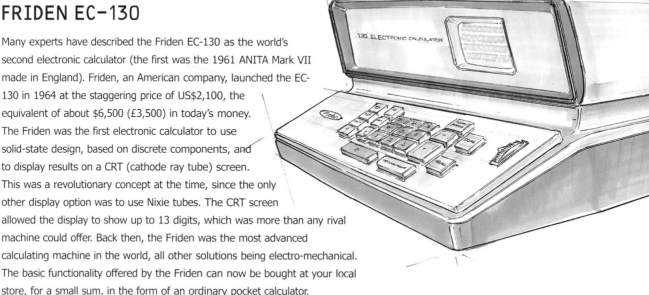

1965
DIGITAL PDP-8

The Digital PDP-8, considered by many to be the first true "minicomputer", was designed to give customers control over programming their own solutions. Before the PDP-8, computing was an extremely expensive option that was available to only the wealthiest corporations who bought packaged solutions. The PDP-8 rapidly found favour in academic circles. Students and staff, who had previously needed to share time on expensive mainframe computers, now had continuous access to computing. The PDP-8 also allowed enthusiastic students to develop the first computer games, a significant trigger to future gaming developments. Although classed as a minicomputer, a PDP-8 is still a considerable chunk of equipment. Even so, it has a large hobbyist following, with an associated market in complete systems and spares.

£500–£2,500 / US$930–$4,650

1970
CANON POCKETRONIC

Although it was called the Pocketronic, this handheld calculator wouldn't fit into any normal pocket. The Pocketronic was the commercial result of research carried out by Texas Instruments and was the first calculator to be based on Large Scale Integrated circuits. It sold for about US$400 (£215). Rather than using an electronic display, results delivery was carried out by a tiny thermal printing mechanism that sent printed paper out of a small slit on the side. This solution was probably chosen to reduce the size and cost. The Japanese-made calculator used 13 rechargeable cells and weighed more than one pound. Texas Instruments was to enter the retail calculator market on its own in later years, becoming quite a force in this competitive arena.

£200–£300 / US$370–$560

1971
RAGEN MICROELECTRONIC

The Ragen Microelectronic pocket calculator was an enigma, since it is quite possible that no examples were ever sold. For such an early calculator, the Ragen Microelectronic promised some astonishing features, such as a Liquid Crystal Display. With a case size of 2.4 x 3.5 x 0.9 inches, it was tiny and used a replaceable 12-volt battery. This would have given it a good service life since the LCD had a low power requirement. The retail price was predicted to be below US$100 (£55) and there was great demand for such a futuristic device, but production issues meant that no working examples were ever seen. Non-working examples were used for marketing purposes and such items are highly prized by calculator collectors.

£35–£60 / US$65–$110

1971
SHARP EL-8

The Sharp EL-8 employed one of the earliest types of electronic digital display: an array of Nixie tubes. Nixie tubes used vacuum tube technology and were essentially complex valves. Each tube contained ten metal electrodes, shaped into digits, arranged one behind the other. The digit-shaped electrodes glowed orange when powered, thus displaying the desired figure through the tube. The drawback was heavy power consumption and large rechargeable batteries were required. The EL-8 originally sold for about US$345 (£185).

1972
CASIO MINI

The Casio Mini was one of the famous Japanese company's earliest models and used miniature fluorescent display tubes. To keep costs down, the first models lacked a floating point and had no decimal point key. The retail price was very low for the time – less than £40 (US$75) – and this helped to make the Mini a very popular calculator. The design was unusual because the display was to the left of the keypad rather than the usual, vertical configuration. It was a very small device, measuring 5.8 x 3 x 1.7 inches, and came with a wrist-strap similar to those used on small cameras. The design set the look of Casio calculators for the rest of the 1970s, during which time the manufacturer would sell millions of units all over the world.

£15–£25 / US$30–$45

£60–£100 / US$110–$185

1972
HEWLETT-PACKARD HP-35

Hewlett-Packard was the first manufacturer to produce a pocket calculator with transcendental functions, using reverse Polish notation. The HP-35 was labelled as the world's first electronic slide rule. Market research suggested that it was too small to sell, but the first year's sales were ten times the forecast figure. In 1972 it sold for US$395 (£210), which was very competitive given the number of functions it offered. The case size was 3.1 x 5.8 x 1.3 inches, so it was more of a handy size than pocket-sized. Build quality was very high for the time, which explains the large number still working and in use today. The HP-35 is a very important part of computing history and highly desirable to collectors.

£120–£220 / US$225–$410

Sinclair Executive

1972
SINCLAIR EXECUTIVE

The Sinclair Executive was Clive Sinclair's first venture into the calculator market and is considered by many as the first truly pocket-sized calculator. The calculators available at the time could hardly be called beautiful, but the Executive changed all that and was described by *New Scientist* as "a piece of personal jewellery". It was even exhibited at the Museum of Modern Art in New York. The display used a Light Emitting Diode (LED) array, which was bright but needed a lot of power. The 2.5-ounce device was launched at a price of £79.95 (US$150). It suffered from a few technical problems, the largest of which was the short battery life. The Executive is probably the scarcest of the Sinclair calculators and market prices reflect this.

OLIVETTI DIVISUMMA 18

Olivetti is famous for its groundbreaking office machinery, such as its revolutionary Valentino typewriter designed by Ettore Sottsass. Olivetti also used designer Mario Bellini, who concentrated on the company's calculating and computing products. His Divisumma range of calculators was unlike anything else. The Divisumma 18 shunned the fashion for LED displays; instead it used a printout as the sole method of delivering results. The construction was of bright yellow plastic covered with a stretched yellow rubber membrane that formed the keys. The Divisumma was an example of product design taking priority over technical function. Olivetti marketed the calculator at the fashion-conscious businessman or -woman rather than nerdy mathematicians. The Divisumma 18 is one of the few electronic items classed as a twentieth-century design icon and nowadays is very difficult to find. Examples with the rubber membrane in good condition are highly prized.

£60–£120 / US$110–$225

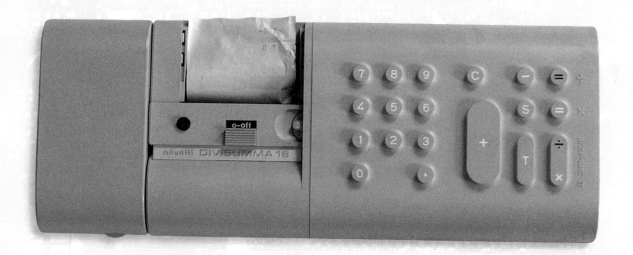

1973
SINCLAIR CAMBRIDGE

Clive Sinclair needed a new calculator to capitalize on the success of the super-slim Executive. In 1973, competition was intense and calculator prices were dropping by the day. His response was the Cambridge. The Cambridge range was sold fully built and in kit form to satisfy the hobbyist market fostered by Sinclair in the 1960s. Despite being thicker than the Executive, the new calculator appeared much smaller at 4.5 x 2.0 x 0.75 inches and, at 3.5 ounces, was slightly heavier. The launch price was £29.95 (US$55) fully built or £24.95 ($45) as a kit, which took about three hours to assemble. More models were added in the following years and included both Scientific and Programmable versions. Collecting all the models in their kit and fully built forms would be a major undertaking on its own.

£20–£60 / US$35–$110

1974

HEWLETT-PACKARD HP-65

The HP-65 was the first fully programmable calculator to use a magnetic card reader/writer. Tiny magnetic cards could be fed through a slot on the side to load or record programs. Hewlett-Packard introduced the calculator as "The Personal Computer", a term that came to be used for describing altogether different machines more than a decade later. Each key on the calculator carried four separate functions, allowing complex programs to be created with a myriad of functions. NASA astronauts used an HP-65 to calculate manoeuvres when making a rendezvous with the Russian Soyuz spacecraft in 1975. The original cost was $795 (£430) and the model was discontinued in 1977.

£100–£250 / US$185–$465

BRAUN ET-23

The Braun ET series of calculators were marketed as lifestyle devices and were designed by Dieter Rams and Dietrich Lubs. Braun was already known for its modern designs of otherwise everyday items, so it was no surprise that the ET calculators were of a high quality. The calculator used a green LED display in a slim case measuring 5.7 x 3 x 0.9 inches. The calculator had only simple functions, the emphasis being on the design rather than what you could do with it. Because it was of interest to only a small market sector, the calculators did not sell in great numbers, which makes them quite a rarity nowadays.

1975

ALTAIR 8800

The Altair, named after a planet in a "Star Trek" episode, was built around a 2MHz Intel 8080 processor with 256 bytes of RAM. It had no keyboard or display: instead, the computer was operated by a series of front panel switches, with results being displayed in a row of flashing lights. It originally cost US$395 (£210) in kit form or $495 (£265) fully built. In the first few months of its life, nearly 4,000 were sold. A version of BASIC was produced for the machine by a new company, then called Micro-Soft, run by Bill Gates and Paul Allen. The Altair 8800 is an important piece of computing history and, as such, is a highly sought-after item by collectors.

£500–£1,500 / US$930–$2,785

£50–£80 / US$95–$150

1975
CALCU-PEN

£100–£150 / US$185–$280

The makers of the Calcu-pen threw away the rulebook when they built a calculator into a writing instrument measuring 6.3 x 0.6 inches. If you can imagine what the impact would be of fitting a mobile phone into a pen today, you'll have some idea of how remarkable the Calcu-pen was in 1975. It had only five buttons but they were cleverly configured to give different functions, depending on which of the four sides was pressed down. The on/off switch was in the pen cap. Its price, in 1975, was just under $80 (£45), which kept the Calcu-pen firmly in luxury goods territory. Good examples are very difficult to find.

1975
CASIO FX-120

The FX-120 used the same green display as most Casio calculators of the time. It had three main modes – deg, rad and grad – and was also configured for statistical functions. One neat feature of this calculator was the ability to convert decimal figures and display them in degrees, minutes and seconds, using a little superscript degree symbol. The high quality of this range of Casio calculators means that while many examples are around today, values are buoyant because of an enthusiastic following of collectors.

£15–£30 / US$30–$55

1975
PRINZTRONIC MINI

Major stores were rushing to cash in on the calculator market and at the front of the queue was Dixons, then (and now) a leading UK electronics retailer. Dixons sold a large range of pocket calculators under its own brand name, Prinztronic. The Prinztronic Mini range was intended to look similar to the phenomenally successful Sinclair Cambridge. It was manufactured in Hong Kong but, contrary to some expectations, its build quality was a match for that of the Sinclair device. These calculators are quite scarce compared to the more common Sinclair competitor yet they have a lower market value, which illustrates well the effect that a famous name can have on prices.

£15–£30 / US$30–$55

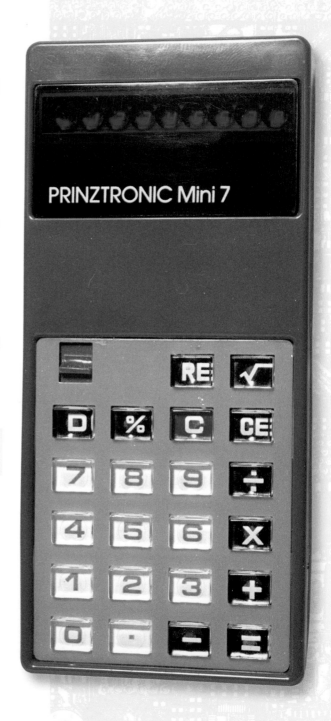

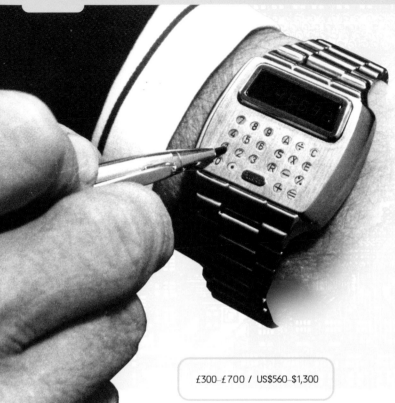

1975
PULSAR CALCULATOR WATCH

In 1975, a calculator built into a watch was in the realms of science fiction – until Pulsar built one. Pulsar made the first commercial calculator watch from solid gold and sold it for US$4,000 (£2,150). It was such a luxury item that the first examples were sold in Tiffany's of New York. The buttons were very small and required the use of a pen or special stylus to operate them. The two large batteries lasted only a few months and contributed to the already hefty weight of the timepiece. In 1976, Pulsar released a stainless steel version for $550 (£295) and by 1977 the price had dropped to $395 (£210), mainly as a result of other makers entering the market. The Pulsar was a high-quality item and many have survived in good condition.

£300–£700 / US$560–$1,300

£20–£60 / US$35–$110

1975
SINCLAIR OXFORD

By 1976, other manufacturers had started to take serious notice of LCD technology and were bringing out calculators using the new display. Sinclair had experimented with LCDs but decided to continue with LED displays. This was a commercially fatal decision that eventually killed off Sinclair's involvement with calculators. The Oxford range was essentially the same as the earlier Cambridge but with a new body. A number of versions were released, including scientific and programmable models. Quality continued to be a problem, particularly with on/off switches, which were prone to failure. The Oxford was a much bigger calculator than the Cambridge and was clearly designed for sitting on a desk.

1976
CASIO CQ-1

Casio was well established in the calculator market by 1976, but other manufacturers were venturing into new arenas and building calculators into everyday items, such as pens and watches. So Casio brought out an innovative range of devices that included the CQ-1, which combined calculator functions with an alarm clock, built into a futuristic desktop case. The "CQ" stood for Computer Quartz and Casio supplied the device in a plush display box to complete the executive look and feel. The CQ-1's design was so good that it wouldn't be out of place on a twenty-first century desk. Build quality was, as usual for Casio, excellent, with the company's trademark green display and good quality buttons.

1976
TSI SPEECH PLUS

The TSI (Telesensory Systems Inc.) Speech Plus was a talking calculator designed primarily for use by blind people. It was based around the ubiquitous Texas Instruments calculator chip but included special circuitry to support a 24-word vocabulary for speech synthesis. The calculator was supplied with instructions on tape and had a special on/off button (on the top right of the calculator in the picture), which also acted as a volume control when pulled out. There was an LED display and a switch for turning off the voice output. Speech was delivered through a tiny speaker located at the back of the calculator and rechargeable cells supplied the power. The Speech Plus would never have sold in large numbers, making it a rare item today.

1977

SINCLAIR PRESIDENT

The Sinclair President is notable for a number of reasons, not least because it was actually manufactured in Hong Kong when most other Sinclair products were made in Europe, and primarily the UK. The President used a power-sapping fluorescent display and was housed in a poor-quality case. It is difficult to come by this calculator today, which shows that sales in the 1970s were poor. Examples have been seen bearing corporate logos, suggesting that they were given away as marketing gifts by various organizations. Despite being perhaps the poorest calculator sold by Sinclair, the President's rarity makes it of interest to collectors.

1977
SINCLAIR SOVEREIGN

It was 1977, the year of the British Queen's Silver Jubilee. Whether this inspired Briton Clive Sinclair is unclear, but this was the year he brought out what many regard as the finest Sinclair calculator ever made. The Sinclair Sovereign was a high-quality instrument housed in a stylish metal case and using LEDs for the display. The silver-plated and engraved version shown in the picture was made to celebrate the Silver Jubilee. Other models were available in chrome and black. Some gold-plated examples were sold and there are rumours that a number were made in solid gold. Possibly the most beautiful calculator ever made.

£100–£300 / US$185–$560

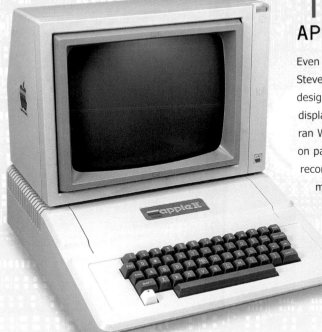

1977
APPLE II

Even while working on the Apple 1, Steve Wozniak, along with Steve Jobs (who was to inspire Apple's revival 20 years later), was designing the next version, the Apple II. The Apple II had a colour display and better memory features than its predecessor. It also ran Wozniak's own version of BASIC, much of which he designed on paper. The Apple II was in production from 1977 until 1993, a record lifespan that's now unlikely to be broken by another single model of computer. Many early computers claim to be the "most important of all time" but the Apple II's contribution to such a vast array of industries, from gaming software to hospital computing, make it one of the top contenders for the crown.

£50–£150 / US$95–$280

£15–£30 / US$30–$55

1977
CASIO FX-2000

Some pioneering manufacturers in the 1970s experimented with Liquid Crystal Display (LCD) technology, but found it unstable and expensive. However, by the late 1970s, LCD displays had become reliable and more economical. The earliest examples had yellow screen filters to protect the display from harmful UV rays. Casio was one of the first to adopt LCD and used it on its high-end scientific calculators, such as the FX-2000. This calculator had all of the features of Casio's larger scientific models but it was a very small, slim and quite an attractive device. The LCD also meant that battery life was nothing less than incredible for the time.

£70–£130 / US$130–$240

1977

COMMODORE PET

The Commodore PET was probably the first computer that worked straight out of the box. It came complete with a monochrome monitor, keyboard and tape storage device. The operating system was built in, so the computer booted directly when switched on. All this sounds fairly obvious but in 1977 it was cutting-edge and contributed to the PET's substantial success. The first models had an awkward keyboard, which may have been attributable to Commodore's calculator background. Later versions had a much better keyboard, but this meant that the tape drive had to be moved to an external housing. The first PETs had a 1MHz processor, 4K of memory and cost US$595 (£320). Later 8K versions cost $795 (£430).

1977
HEWLETT–PACKARD HP–01

Other manufacturers, such as Pulsar, had already introduced watches with built-in calculators but this didn't deter HP from building one of its own. The result was the HP-01, an exquisite calculator wristwatch that has never been surpassed. The HP-01 was expensive at US$850 (£455) for the gold version or $695 (£375) for the stainless-steel one. It was a technical marvel, with a built-in 200-year calendar and the ability to perform calculations on time itself. The construction was of the highest quality, right down to the tiny stylus built into the band's buckle, which was used to operate the recessed buttons. The watch also came with a pen that had a second stylus for key operations, a case for spare batteries and case seals. Examples with the full accessory kit are scarce and the most valuable.

1977
SHARP EL-8130

Sharp's EL-8130 calculator from 1977 was revolutionary in a number of ways. It was one of the thinnest calculators available, at just 0.2 inches deep, and also introduced a flat membrane keypad with no discrete keys. Although a touch keypad had been used on a calculator before, the EL-8130 was the first to use musical tones to denote keys being pressed. The key with the musical note symbol could be used to switch the tone on and off. The display was LCD, which by now had a clear filter showing the grey background we are familiar with today. The construction was high quality, the main body being made from a solid piece of aluminium.

1977
SINCLAIR WRIST CALCULATOR

The truth about the Sinclair Wrist Calculator is that it was a terrible product. It comprised a mix of old calculator chips, an LED display and batteries crammed together in a tiny plastic case. It was sold in kit form and was almost impossible to build because of the variance in component sizes. Remarkably it did sell, with more than 10,000 kits bought, mainly by Sinclair's loyal hobbyist followers. However, sales in the USA were very poor and a good number of kits were returned to the UK. How many were eventually successfully built is not known, but working examples are incredibly rare, as are examples of the actual kit. This, unsurprisingly, is one of the rarest of Sinclair's products.

£100–£250 / US$185–$465

£30–£60 / US$55–$110

1977
COMMODORE S61 STATISTICIAN

The Commodore S61 Statistician must hold the record for the most keys on a pocket calculator – 60 in total, many of which had more than one function. It had a green fluorescent display that allowed ten digits mantissa and two digits exponent. It was quite rare for calculators to be dedicated to statistical functions and the S61 may have been the first example. It could be held in one hand, but its size (6.2 x 4 x 1.3 inches) made it more suitable for desktop work. Being a specialist calculator meant that not many were sold, so it is a rare find today.

1978
SINCLAIR PDM35 MULTIMETER

£15–£30 / US$30–$55

Sinclair's products included workshop instruments, such as multimeters and even an oscilloscope that used a tiny cathode ray tube. The PDM35 was the smallest of the Sinclair multimeters but was still a very precise measuring instrument, with one percent accuracy and a three-digit display. Sinclair launched the PDM35 in the USA for a price of US$49.95 (£25). The similarity with the Oxford range of calculators was not coincidental since the PDM35 was housed in the same case moulding. It was a well-made device and most examples seen today are in working order.

£40–£80 / US$75–$150

1978

SINCLAIR ENTERPRISE PROGRAMMABLE

The Enterprise Programmable was Sinclair's last attempt to recapture success in the calculator market. It failed – and sold at a loss because it could not compete with Japanese-made LCD calculators, which were simply better machines. The Enterprise's LED display immediately marked the calculator as being "old technology" and possibly stifled sales. It was sold with a set of program libraries ready to use out of the box. It was actually a nicely made device and compact at 2.6 x 5.3 x 0.9 inches, but it was difficult to change the battery, which required the removal of the whole top of the case.

1980
SHARP MZ80K

After the Commodore PET's success as an "out of the box" computer, other manufacturers launched their own models. One of the earliest was Sharp's MZ80K, the "Z80" referring to the processor used. It was first sold in kit form, but by late 1979, it was a fully built, ready-to-run computer. The Sharp came with a peculiar calculator-like keyboard similar to the one on the Commodore PET, but it proved unpopular and was replaced in later models. Data storage was provided by a built-in tape recorder. The Z80 processor ran at 2MHz. Basic MZ80Ks were supplied with 20KB of RAM and a 10-inch monochrome monitor.

£80–£150 / US$150–$280

1980
SHARP PC-1211

The Sharp PC-1211 (also sold as the Radio Shack TRS-80 PC1) can probably claim to be the first true pocket computer, not least because it incorporated a full QWERTY keyboard in a 6.9 x 2.8 x 0.7 inch case. Results were displayed on a 24-digit alphanumeric LCD panel. Note the yellow screen filter, which indicates an early LCD display. The PC-1211 incorporated the BASIC programming language, which could be used for 1,424 steps using up to 178 memories. Available accessories included a tape/printer interface that made the PC-1211 a truly useful business computer. The batteries were good for 300 hours' use, which was incredible for the time.

£50–£100 / US$95–$185

1980
RADIO SHACK TRS80 1

£50–£200 / US$95–$370

In 1977 Radio Shack launched its range of computers with the TRS80 1, priced at US$400 (£215). The computer circuitry was incorporated into the keyboard housing, a design that would catch on with other computer manufacturers in years to come. It was possible to buy the computer as a complete package, with a monitor and cables, so that you could start computing right away. Like many early computers, the TRS80 came with the BASIC programming language on board. The model shown in the photo dates from 1980 and has a floppy disk drive – an expensive storage option at the time. The highly successful TRS80 1 remained in production until 1981. During that time, two further TRS80 models and a colour version were released.

1980

INTERTEC SUPERBRAIN

The Intertec Data Systems Superbrain was a substantial machine, weighing in at 45 pounds. It certainly lived up to the futuristic expectations people had of computers in 1980 and, at a casual glance, wouldn't look out of date today. It was particularly interesting for its use of two Zilog Z80 processors, each running at 4MHz. One was used to control the two 320K floppy disk drives while the other handled everything else. There was 64K of RAM and an RS-232 port on the back of the case for peripheral connection. The operating system was CP/M, which was an excellent system for controlling computers. The original cost of the Superbrain was US$3,000 (£1,615), putting it firmly in big-business territory.

£150 / US$280

1981

£30–£80 / US$55–$150

IBM PERSONAL COMPUTER

IBM had been watching the growing personal computer market for some time when, in 1980, it called on a young Bill Gates to write software for a secret project. IBM had already tried and failed to break into the personal computing market, so, for its second attempt, it wanted a brand-new operating system to go with the new hardware. The result was the IBM Personal Computer, based around the 8088 CPU with 16K of RAM and running the new Disk Operating System (DOS). There was one floppy disk drive (or two, if you could afford the upgrade). A colour monitor was available but was an expensive option. The original price ranged from US$3,000 to $6000 (£1,615–£3,225). Perhaps the most important aspect of the IBM PC was that it was built from "off-the-shelf" components, a concept that evolved into the modern PC as we know it.

1981

OSBORNE PORTABLE COMPUTER

The Osborne 1 was the first successful portable computer. Everything you needed for computing was included in a briefcase-sized box. The keyboard was housed in the lid, under which was the carrying handle. There were two floppy disk drives and a tiny, five-inch, monochrome monitor. The little screen was capable of displaying only 52 characters on a line, so to see the complete text, you had to use a special scroll function that moved the display back and forth. Osborne chose the CP/M operating system for the computer, which was sold for US$1795 (£965). The Osborne was a massive success and sales hit 10,000 a month. Unfortunately the company went under in 1983 as the public waited for better Osborne machines that never materialized.

1982

HEWLETT-PACKARD HP16C

Hewlett-Packard launched its famous HP16C calculator in 1982, the company's first calculator with functions specifically designed for computer programmers. The calculator was extremely well built and used a modern LCD screen. The power was supplied by three tiny button cells, which allowed the case to be very small and slim at 5.1 x 3.1 x 0.6 inches. The available functions were extensive and included hexadecimal conversion and adjustable word size, as well as integer and floating point arithmetic. The original cost was a hefty US$150 (£80) but the HP16c was very successful and remained in production until 1989. It is in great demand today by both collectors and buyers who want to use it in their work.

1982
EPSON HX-20

£30–£100 / US$55–$185

Epson was better known as a printer manufacturer and supplier of LCDs, but it surprised the computing market in 1982 by launching the remarkable HX-20 portable computer. The A4-sized (a standard UK paper trim) HX-20 came with a full-sized keyboard, built-in tape storage, a printer and an onboard battery that gave 50 hours of use. It can be regarded as the world's first true laptop computer. It was also 20 pounds lighter than the Osborne 1, the only other serious portable computer available at the time. To achieve a package so small, some compromises were made. The processor ran at a very slow 0.6MHz and the screen was quite a small LCD unit capable of four lines of 20 characters.

1982
RESEARCH MACHINES RM-380Z

Like many early 1980s computers, the RM-380Z was based around a Z80 processor with 32K of RAM. Aimed at the professional computing market, it became popular in schools. The computer was housed in a substantial case (23.4 x 16.7 x 8.4 inches) with two eight-inch floppy disk drives. There was a separate keyboard and a choice of monochrome or colour monitors. The operating system was initially a bespoke version of CP/M, but this was later standardized to allow the use of other software. The 1982 price was £2,000 (US$3,715) upwards.

1983
CASIO PB-700

Casio, already a very successful maker of pocket calculators and digital watches, had not been ignoring the growing market for pocket computers. At the time, Sharp was dominant but then Casio stepped in with its brilliant PB-700. For a pocket computer, the screen was very large and offered four lines of 20 characters. It had a full QWERTY keyboard, 4K of RAM, built-in BASIC and a little speaker for simple beep sounds. The case was sleek and compact at 7.9 x 3.4 x 0.9 inches. The PB-700 could be expanded with extra RAM modules and could be connected to an external tape recorder. The 1983 price was about £350 (US$645) so it was expensive, but the large number sold meant it was a very popular machine.

1983
APPLE LISA

£300–£600 / US$560–$1,110

Xerox was the first computer manufacturer to use a Graphical User Interface (GUI) and a mouse on its computers, but they were not a commercial success. The Apple Lisa offered both a GUI and a mouse in a business package for less than US$10,000 (£5,415). Based around a Motorola 68000 5MHz CPU and with just 1MB of RAM, the Lisa was hardly powerful by modern standards, but it was considered a very advanced machine in 1993. A huge amount of research went into bringing the Lisa to the market, yet, despite this, it was not a commercial success – not least because Apple was soon to launch an entirely new, ground-breaking machine called the Macintosh.

1983
CANON X07

The Canon X07 took portable computing to a new level. Not quite pocket-sized, the X07 was still very compact at 7.9 x 5.1 x 1 inches. Despite its diminutive dimensions, it still had a QWERTY keyboard (albeit with calculator-style keys), a four-line LCD display, sound, memory expansion slots and both serial and parallel ports. It also had the ability to communicate using infrared, which was a very advanced concept in 1983. The fact that the X07 could be made to interface easily with other equipment made it a very popular scientific computer. It wasn't cheap at a launch price of just under £400 (US$740) and is a very rare find today.

1983
SEIKO DATA 2000

What could you buy the 1983 businessman who had everything? How about a wristwatch with a built-in data bank? Seiko was the market leader in LCD digital watches and was experimenting by adding new applications to its wrist-worn technology. The Data 2000 (the "2000" is derived from the watch's 2KB of memory) came with a keyboard on which the watch could rest. The keyboard communicated with the watch using electrical induction and allowed full editing of the watch's data. The screen employed a dot-matrix LCD rather than the segment-based screens more commonly seen on watches. This allowed great versatility in character and graphic display. Seiko would later bring out versions that could be programmed in BASIC.

1984

HEWLETT-PACKARD HP-71B

Like Hewlett-Packard's extensive range of calculators, the HP-71B was a beautifully constructed machine. Clearly, extensive care and thought had gone into the hardware design as well as the software. The screen was a single-line 22-character LCD and you could scroll the display to view five lines and 96 characters. The full QWERTY keyboard used the same high-quality keys as previous HP machines. Memory was a generous 64K, which could be expanded using modules. The HP-71B could be enhanced with the addition of a card reader and connected to peripherals using the HP-IL interface standard.

1984

IBM PORTABLE COMPUTER

By 1984, portable computers were becoming readily available. Compaq had entered the market with a competitive IBM-compatible machine and IBM's offering was the 5155 portable PC. IBM had made a portable computer in the 1970s, but that was a huge machine with an equally huge price. The 5155 was priced at under US$4,500 (£2,440). The term "portable" was just about justifiable for a 30-pound machine that had to have a power supply to work. The nine-inch amber screen was good quality, as was the keyboard, and the heart of the machine was the same as the already successful IBM desktop PC.

1984
APPLE MACINTOSH CLASSIC

The Apple Macintosh was not the first machine to offer a Graphical User Interface and mouse, but it was the first *affordable* personal computer to use them and it sold in high numbers to both commercial organizations and the general public. The Macintosh had a processor speed of 8MHz, 0.5 MB of memory and was the first personal computer to use the new 3.5-inch floppy disks. All this for US$2,495 (£1,350). The Classic is seen as a design icon that has only recently been surpassed in the computing world by the flat-screened Apple iMac range. The later Classics are still usable computers and examples are easily found at boot sales and computer fairs.

£30–£80 / US$55–$150

1984
PSION ORGANISER

Psion launched its first organizer in 1984. Called the Psion 1, it sold for £99 (US$185). With 2KB of RAM and an LCD display, the Psion was an immediate success. Later models introduced larger displays, more memory and better application support. The Psion models were incredibly power-efficient and could run for six months of use on a single 9-volt battery. Applications were available from third-party companies in the form of ROM packs that would slot into the backs of the organizers. Psion machines, particularly later models, were very popular with financial companies, who would issue them to their staff for loan and mortgage calculations.

£15–£60 / US$30–$110

1984

HEWLETT-PACKARD HP 110

Hewlett-Packard's first portable MS-DOS computer cost just under US$3,000 (£1,615) in 1984. It had a remarkable specification for the time, with a 5MHz CPU, 256K of RAM, built-in Lotus 1-2-3 software and even a modem. Compared to its contemporaries, it was relatively light at 9 pounds, most of which was the battery, which provided up to 16 hours of use. The screen employed an 80-character 16-line LCD that, for a while, was one of the largest LCDs on a portable machine. The HP 110 received good reviews and, within a year, HP launched a new model that had a larger screen, more memory and a faster modem.

£15–£50 / US$30–$95

1984

MOTOROLA 8500X CELL PHONE

In 1984, Motorola launched the first cellular mobile telephone service. It could have been sooner but for FCC rules holding up the release of frequencies. The first real hand-held mobile phones were part of the brick-like Motorola family, of which the 8500x was a member. Early phones, such as the 8500x, were analogue only, but a few later models of this shape worked with the digital GSM system. These are now highly sought after, despite the fact that they are several times the size of modern equipment, because they still work on today's networks.

£40–£100 / US$75–$185

£50–£150 / US$95–$280

1984
SINCLAIR QL

Clive Sinclair wanted to launch a serious computer for office use. The result was the Sinclair QL (the "QL" standing for Quantum Leap). It had a stunning appearance, in marked contrast to the other uniformly grey machines on the market. The QL was in high demand and advance orders reached 13,000 just months after the launch, surprising Sinclair's production, which failed to match the demand. The QL ran a special operating system called QDOS and had rapidly accessible, removable storage in the form of Microdrives (small tape-loop cartridges). The keyboard looked very modern, but proved difficult to use for natural typists. To enforce the business orientation of the QL, Sinclair refused to produce games software for the machine, a decision that may have accelerated the eventual downfall of the sleek black box.

£20–£50 / US$35–$95

1985
AMSTRAD PCW 8256

Amstrad, a company famous for bringing out very cheap products loaded with functionality, launched the Z80A-based PCW (Personal Computer Word processor) with a price of £460 (US$850). This was remarkable for a machine that included a dot-matrix printer, operating system with software and two programming languages, BASIC and Dr Logo. The PCW was marketed primarily as a word processing solution but it was actually a useful computer running the CP/M operating system. Everything worked via a single power cord and data was stored on compact three-inch floppy disks. Later models came with more memory, better screens and a daisy-wheel printer. It sold in huge numbers and still has an enthusiastic band of followers.

1988
CAMBRIDGE COMPUTERS Z88

Clive Sinclair's Z88 laptop should have been a much more successful machine. It was highly portable at less than two pounds and just 11.6 x 8.2 x 1 inches, had an excellent keyboard and ran for many hours on four AA batteries. It was built around the Zilog Z80 processor and used a wide LCD display, which allowed eight lines of 104 characters. A version of BBC BASIC was built in, along with a word processing program and a diary system. Transfer of data between the Z88 and a PC was possible using a cable. Expansion was available for both memory and EPROM modules.

£40–£100 / US$75–$185

1989
ATARI PORTFOLIO

£40–£80 / US$75–$150

The Atari Portfolio can rightly claim to be the world's first truly pocket PC that ran a real operating system, in this case MS-DOS. It even had a usable keyboard, which is more than many of today's pocket PCs can offer. It was very popular, not least because of its US$399.95 (£215) price. Being only 7.5 x 4 x 2.5 inches, it was very compact and weighed just over a pound. At its heart was an Intel 80C88 CPU running at just under 5MHz, which was as fast as IBM's first PC (compare that with today's laptop PCs, which run up to 600 times faster). The display allowed for 40 characters over eight lines. There was even a tiny speaker and the ability to interface with other computers and networks. The Portfolio found fame in *Terminator 2*, where it was used by the young John Connor to hack into cash machines.

HEWLETT-PACKARD HP 95LX

The HP 95LX was Hewlett-Packard's first palmtop computer. At the 95LX's launch, HP claimed that it had as much computing power as a desktop personal computing system. It had MS-DOS in ROM and built-in software, including a telephone/address book and the Lotus 1-2-3 spreadsheet. Some experts consider the 95LX to be the first Personal Digital Assistant (PDA). It was available with up to 1MB of RAM and an infrared port that was powerful enough to be used as a TV remote control. As one would expect for an HP machine, the 95LX was built to a very high standard using the best materials. At just 12 ounces, the robust and light HP 95LX set the standard for palmtop computers.

£20–£50 / US$35–$95

£30–£130 / US$55–$240

PSION 3

Psion's success with its original Organiser led to the introduction of a new type of pocket computer with a new operating system. It was called the Psion 3 and became a phenomenal success. A number of other models followed, with larger screens and back-lit displays. These computers used a clamshell design, had a usable keyboard and allowed for memory expansion or ROM-based programs to be run. The Psion 3MX was the last in the 3 series of pocket computers and delivered a threefold increase in processor speed. The Psion 3 MX is the most popular model with collectors.

1992
SWATCH THE BEEP

Other manufacturers had been building calculators and
even televisions into their watches but Swatch came up
with the brilliant idea of building a pager into one of its
timepieces. In the early days of cell phones, many people still
used radio pagers since they were much cheaper. The Beep kept
Swatch's preference for an analogue time display but included a digital
display, which was necessary to show a paged message. Up to ten
messages could be stored in the watch, which cleverly had the pager
aerial built into the crystal. The Beep was larger than a normal Swatch
watch and suffered from poor battery life.

1993
APPLE NEWTON

Apple took a giant leap in 1993 when it launched the Newton, a
tablet-style personal computing device with handwriting
recognition. It had a large, reflective LCD display, which could
be operated using a stylus. The Newton was large (7.3 x 4.5 x
0.7 inches), very expensive (more than US$600/£320) and, at
just under one pound in weight, too bulky for the largest pocket.
Throw in handwriting software that didn't work well, and it just
didn't hit the mark. Later models improved on the handwriting
software and had redesigned cases, but they came too late. By
then, the market for PDAs had been won by Palm and Pocket PC
devices. The Newton was made until 1997.

1996

PALM PILOT PERSONAL

The Palm Pilot revolutionized the way we carry information around with us. Before the 1MB of memory device appeared, the market for personal information management was the preserve of the Filofax. There had been attempts to produce a portable information device before, but it was the Palm's amazing "Graffiti" method of writing recognition that dispensed with the need for a keyboard and allowed the handy format of the organizer. The PDA market is one of the fastest moving today and products are out of date within months of their launch. Hardly anyone uses the first Palm Pilot model now, which makes it of great interest to collectors.

1997

PSION 5

By the second half of the 1990s, the market for personal computing was becoming crowded with emerging Palm and Pocket PC devices. To keep pace, Psion launched the Psion 5, a machine that retained the clamshell and keyboard format but improved on the Psion 3 in all aspects. The keyboard was much better, folded in a more sophisticated way and had a real keyboard feel. The screen was larger, back-lit and could be operated with the provided stylus, which slotted neatly into the side of the case. The operating system was still Epoc, but now a 32-bit version running on a faster processor with 4MB or 8MB of memory available. In 1997, Psion released the 5MX, a more powerful machine and the one most in demand today.

1998
APPLE IMAC

The Apple iMac was a shot in the arm for a computer industry stuck in a beige box rut. Not since the 1984 Apple Macintosh had anything so bold and different been seen. Apple's sales were floundering and the company needed to recapture a share of a market then dominated by Windows-based PCs. The iMac was aimed at low-end consumers, with a colourful design, affordable price (under US$1,300/£705) and built-in Internet access. The various attractive colours were highlighted by one of Apple's marketing slogans: "Sorry, no beige...". The iMac became the fastest-selling computer in history.

£150–£300 / US$280–$560

LG PHENOM

Other manufacturers were using Microsoft's CE mobile operating system for compact PDAs, but a small number of companies were using it for keyboard-centric palmtop devices. In the late 1990s LG introduced the Phenom range of machines, which used Windows CE and featured a full-sized high-quality keyboard. The "instant-on" capability of the machine made it very attractive to those wanting to capture text on the move. The 8.5-inch colour screen was touch-sensitive and a stylus could be used to operate the applications. The Phenom had a printer port, a standard VGA output, PCMCIA slot and infrared capability, so it was a very useful machine. A built-in LiIon battery gave it reasonable battery life, but it couldn't compete with the longer battery.

£50–£130 / US$95–$240

£30–£60 / US$55–$110

£70–£110 / US$130–$205

2002
OLYMPUS DM-1 VOICE RECORDER

The ever-falling price of memory meant that it no longer had to be confined to computers. Voice recorders and music players were an obvious choice – and clearly a successful choice, given today's proliferation of digital voice recorders and MP3 players. One of the first serious voice recorders was the DM-1, which was also a digital music player. It could be connected to a computer using the fast USB protocol and had a back-lit screen for easy use at night. The DM-1 allowed the use of removable memory cards. A 128MB card could store up to 45 hours of recordings, which could be organized into folders for easy retrieval. A well-built device and a good bet for a future collector's item.

2000
PALM PILOT IIIC

The Palm Pilot IIIc was the first version of the Palm range to have a colour screen. By 2000, there was fierce competition from an increasing number of personal machines that used the up-and-coming Windows CE operating system. Palm still had the edge, however, since the IIIc was smaller than any of the existing colour Windows CE products. It had 8MB of memory, which was competitive at that time. It should have taken the market by storm, but it didn't prove popular with Palm customers and was finally dealt a deathblow when Palm introduced the much smaller and sleeker M505 colour Pilot in 2001.

2002
SONY VAIO PCG-U3 NOTEBOOK

£500–£800 / US$930–$1,475

The PCG-U3 notebook was designed for the computer user who was constantly on the move. This tiny machine measured just 7.3 x 5.5 x 1.2 inches and weighed less than two pounds, but it included a 933MHz processor, 512MB of RAM and a 20GB hard drive. Despite being just over six inches across, the diminutive screen could resolve at up to 1024 x 768 pixels, with 16 million colours. The PCG-U3 was basically a fully featured, full-power PC in a package the size of a large PDA. Of course, the drawback of such a package was the limited ability to upgrade.

2002
SHARP ZAURUS C750

Sharp's Zaurus range of PDAs were peculiar in that they ran the Linux operating system, beloved of many computer technophiles. For that reason alone, they became a highly desirable "techie toy". Loaded with all the functionality of other PDAs, the Zaurus also had that extra air of exclusivity: membership of the Linux club. The Zaurus was very well built and because it was available mainly in the Far East, Western second-hand prices can be quite high.

£200–£300 / US$370–$560

2002
BLACKBERRY

Blackberry launched its "always on mail" service in late 1999 and the idea caught on in a big way. What Blackberry offered was the ability for corporations to deliver mail to and from their enterprise servers to the remote Blackberry devices used by their staff. The ability to remain in touch with your staff 24/7 was too much for businesses to ignore, so Blackberry sales took off. The devices were large, cumbersome and not pretty, but they became *the* gadget to have. Blackberry brought more models into the market and launched the idea in Europe in 2002, offering the service through existing operators to make the Blackberry device more attractive to private individuals.

£80–£150 / US$150–$280

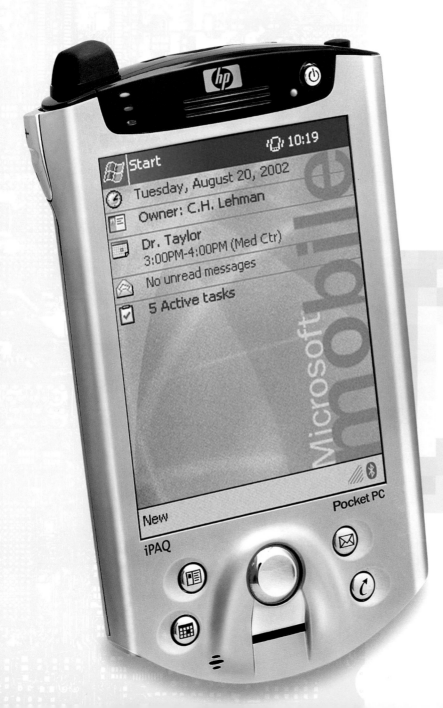

2003
HEWLETT-PACKARD
IPAQ

The iPAQ pocket PC was devised and introduced by Compaq back in 2000. It was an innovative product at the time, running Microsoft's Pocket PC operating system. Two major selling points were the very large colour screen and futuristic chrome case. However, it suffered some problems, not least very short battery life and dust that got into the screen. When Compaq was merged with Hewlett-Packard, the iPAQ lived on as an HP product. The model shown here was a considerable improvement on the original, with a better and dust-free screen, more memory, more power and even the ability to read a fingerprint to secure data.

£100–£300 / US$185–$560

2003
SONY CLIE PEG-NR70V

While other big corporations were backing Microsoft's Pocket PC operating system, Sony decided to run with the Palm platform. Of course, Sony did it in style and its first offerings were high-quality jewel-like devices. The Clié PEG-NR70V was its top machine when launched, at a price of just under US$600 (£320). It used a Motorola 66MHz processor and had 16MB of RAM. The screen was large for this type of device and could be rotated at the base. The Clié was fitted with a low-resolution colour camera and a tiny keyboard, which was really too small to be of much use. The high price and Sony's proprietary memory solution may have put many potential buyers off, making this quite a scarce machine, even though it was made relatively recently.

£80–£120 / US$150–$225

WORKSTATION...

57

2004
APPLE IBOOK G4

The Apple iBook G4 has to be the most
beautiful laptop computer yet made. Many PC-
based laptops have tried to emulate the
iBook's looks but none has come close. The
iBook was made in two sizes, but both kept
the same sleek appearance. The big machine
had a wide, 14.1-inch colour display driven by
a powerful graphics chip. One of the key
selling points was the incredible battery life,
which was up to twice that of the best laptop
PCs from other manufacturers. The iBook ran
Apple's own Mac OS operating system, which
limited the machine to Mac enthusiasts, but
the iBook was a massive success.

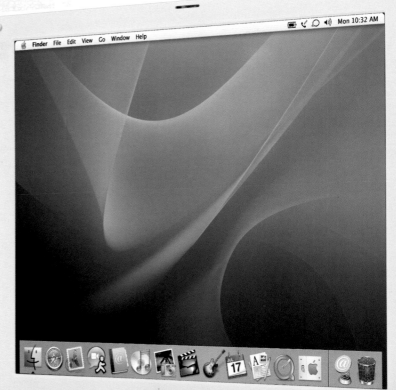

£500–£800 / US$930–$1,475

£150–£200 / US$280–$370

2004
HANDSPRING TREO SMARTPHONE

By 2004, smartphones were here to stay. Product convergence meant that these little technical marvels could be used as a cell phone, a PDA, a digital camera, a music player and, in many cases, an e-mail and web access device. The Treo was a PalmOS-based machine, which opened it up to the many thousands of Palm applications available on the web. The built-in memory could be expanded using memory cards and it even had a stylus for screen operation, although it lacked the handwriting recognition of the dedicated Palm PDAs. Battery life was incredible and allowed the Treo to run for many days between charges.

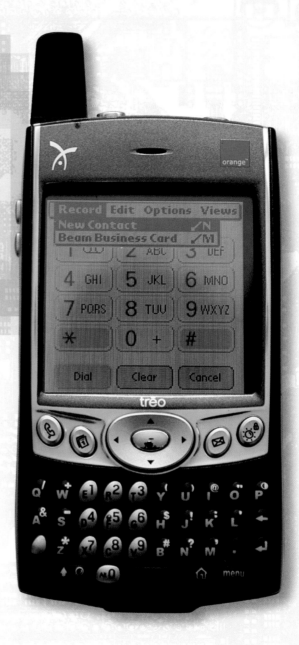

HOME BASE

It feels as though waves of new technology are constantly attacking our homes. We are under pressure, like never before, to buy the latest and greatest "digital" product available. For today's home owner, the world is a very confusing place. Yet it hasn't always been like this. As recently as the 1960s in the UK, the average home would have been relatively free of any technical products, other than a radio receiver or, for wealthier families, a television set. Halcyon times, when life was slower and "stress" was a rarely heard word in the home.

In those early days, the first technical advances were limited to entertainment devices that were either radios, record players or, rarely, a reel-to-reel tape recorder. Then in the early 60s a new medium arrived in the form of the Compact Cassette, which allowed personal recordings to be carried around easily and would remain popular for nearly 25 years. The lava lamp entertained the hippy generation and famous designers made humdrum equipment trendy. However, none of this could prepare households for what was to come in the next 20 years.

The 1970s brought with them an increase in the rate at which technology invaded our homes. This was the decade of knobs and lights. The more lights, switches and controls a product had, then the better, more expensive and cooler it was. The flashy trend stayed with technology well into the next decade, but some manufacturers eschewed it to introduce the world to minimalist cool, albeit at a price. The biggest impact to home life was, perhaps, the availability of affordable video recorders, which instantly changed the viewing habits of millions. Or perhaps it was the arrival of a tiny tape player called the Sony Walkman – soon to become the first personal-technology icon.

In the 1980s, we still had the lights and knobs, but now the world wanted black instead of the brushed aluminium of the previous decade. The public's hunger for new gadgets was fuelled by increased disposable income and the arrival of the "yuppie" generation. A new medium for music arrived: the Compact Disc. With vinyl records firmly entrenched as the method of playing back music, very few people would have bet on the digital upstart becoming

  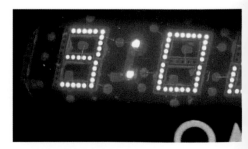

the medium of choice. Looking back, it's hard to believe that it was ever a new technology, and for a while we thought it would last forever. Video cameras became popular, starting off as suitcase-sized behemoths but becoming small enough to fit in the palm of a hand by the end of the decade – by which time nearly everyone owned a personal music device inspired by Sony's original Walkman.

The onslaught continued into the 1990s. Digital still cameras appeared, albeit with very low resolution at first and no immediate threat to the film industry. However, the improvement in quality was dramatic and digital cameras became cheap and efficient by the end of the decade. The Internet became a household word and, for many, it was soon an indispensable part of daily life. The web brought with it online auctions, destined to change the way we shop, as well as digital music that would soon threaten sales of Compact Discs. Portable players for digital music became available and the Apple iPod became etched into consumers' consciences overnight – bringing with it a threat to the dominance of the CD as downloadable music became available.

Here in the new century, we are seeing more life-changing technology than ever before. Recordable DVD players are killing off video recorders. Downloadable music is becoming the medium of choice and the digital camera now outsells film cameras, changing forever the way we record memories. Convergence is the path products are following. Our telephones are also cameras and music players, and are Internet-connected. Our digital flat-screen televisions can display photographs and playback movies with perfect surround-sound. Digital music is streamed to every room in our homes through wireless networks and news reaches us instantly via the Internet. Most people can be contacted anywhere at any time by voice, mail or text. What our parents considered pure science fiction is, to us, normal and unremarkable.

The home is where it is all happening, whether we like it or not. Technology is there to enrich our lives. Don't resist it: use it, enjoy it, collect it!

1962
BULOVA ACCUTRON

The look of the Bulova Accutron watch
makes it hard to believe that it was
designed nearly half a century ago. The
Accutron had a tiny electronic circuit, which
used a tuning fork for time regulation.
Tuning forks had already been used on
larger-scale electric clocks, but miniaturizing
the idea for a watch was a remarkable feat
at the time. The Accutron emitted a
beautiful, high-pitched hum that was
generated by the tuning fork, which was
vibrated electronically. Accutrons were made
for many years and in many styles, but the
Spaceview model shown here, dating from
1962, is by far the most sought after.

£120–£450 / US$225–$835

1962
SONY TR650 RADIO

Sony had enjoyed a good reputation for "shirt-pocket" radios since the 1950s. Its TR650 radio still had some 1950s features – especially the large, round speaker grille – but it was destined to be the last of that phase. This small radio was very successful and proved particularly popular with young buyers who wanted to hear their new music on the move. Other companies, from Japan and other countries, copied Sony's radios but despite their popularity, sales back then never approached today's tens of millions. As a result, this little Sony radio is quite a scarce and an expensive item.

1963
AMPEX VR-660B VIDEO RECORDER

Not many people realise that video tape recorders existed as far back as the late 1950s. This 1963 Ampex VR-660B machine was a fully transistorized, broadcast-quality portable recorder, pitched at the professional market, and priced at US$14,500 (£7,825). In its day, that was very cheap for a broadcast-quality machine. It used a helical scan mechanism, recording on to two-inch tape. A single tape reel could record up to five hours of continuous play. It was hardly a consumer item at that kind of price so relatively few were made, which makes it a tantalizing technology item for collectors today.

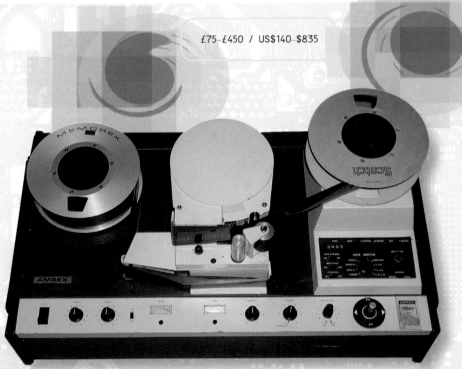

1964
BRIONVEGA ALGOL TELEVISION

Designed by Marco Zanuso and Richard Sapper, the Algol portable television was years ahead of its time – so much so that Brionvega makes a very similar television with the same name today, albeit with updated electronics. Brionvega was known for using the best designers. The company targeted its products at "those who appreciate style and design, and who expect something different, more flamboyant". The Algol can be seen in design museums all around the world so it's unsurprising to learn that it is highly attractive to collectors.

1964
BRIONVEGA TS502 RADIO

Zanuso and Sapper, the same pair of designers responsible for the Algol television (above), designed this smart cube radio for the stylish Italian manufacturer. The TS502 – not such a stylish name – was a transistorized radio that opened on a hinge to reveal the speaker and controls. The radio had AM/FM reception and could work when closed. Note the control indicators, which have the appearance of a car dashboard. The TS502 was another of Brionvega's products with a timeless design, making it an extremely desirable collector's item today.

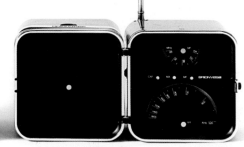

1964
LAVA LAMP

Craven Walker brought us the Lava Lamp in the 1960s. The idea was simple: a bottle filled with water and a special oil that, when heated and illuminated by a lamp from below, would animate in a globular fashion. You either loved or hated the Lava Lamp and Walker initially had problems marketing the idea, but it became a success as the 1960s adopted the pop and drug culture. The eccentric Walker said of his invention, "If you buy my lamp, you won't need to buy drugs." The lamp shown here is a contemporary model but the design has changed little over the years.

£10–£50 / US$20–$95

1965
SINCLAIR MICRO FM RADIO

Sir Clive Sinclair's first tiny radio appeared in 1963. It was sold as a kit, based on a project he had published in *Practical Wireless* in 1958. His second miniature radio was this Micro FM model. It was also sold as a kit for hobbyists to build. The case was made with a smart, polished aluminium facia and a similar tuning knob. It was relatively expensive, selling at just under £6 (US$11). It reportedly did not work very well, which led to poor sales. Interestingly, clones of the Micro FM were produced in the Far East, giving the little radio a certain cachet.

£100–£200 / US$185–$370

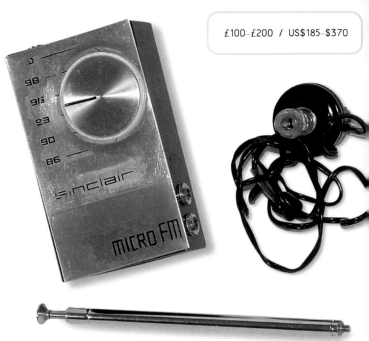

1965
BRIONVEGA RR126 MUSIC CENTRE

Designed by Achille Castiglioni for Brionvega, the RR126 music centre was unlike anything seen before. It was built around a steel frame and used aluminium, Masonite and plastic laminate for decoration. The speakers could be detached and placed on top of the unit, and the controls are organized in such a way that they gave the stereo the appearance of a robot's face. The RR126, which was made in various colours and finishes, comprised a record deck, amplifier and AM/FM radio. The whole unit could be moved around easily using the castors on the steel "foot".

£800–£1,200 / US$1,475–$2,225

1965
ERICOFON

The original Ericofon was made by Swedish company Ericsson from the late 1940s onwards. Compact and streamlined, the Ericofon was originally sold in Europe. It did not meet with much success in the USA, mainly because of Bell Telephone's control of the telephone market at the time. The earliest models had dials, but this 1960s model, sold in the USA by North Electric, used the standard American 12-button keypad. The Ericofon was available in 18 colours. Collecting the complete range of colours and models would probably be a lifetime's work.

£20–£40 / US$35–$75

£30–£60 / US$55–$110

1966
TRIMPHONE

In the 1960s, the UK telephone services were controlled by the Post Office. The Post Office wanted a "luxury" telephone to add to its range and, as a result, commissioned Standard Telephones and Cables to produce the Trimphone. The new telephone design was seen as "space age". It was *the* phone to have and found its way into the smartest homes. The ringer had no bell; instead an electronic oscillator produced a warble. The dial was luminous, glowing green in the dark, which was accomplished using tritium gas in a small glass tube, producing beta radiation that caused the dial to fluoresce. This feature was later withdrawn due to public concern.

1967
SINCLAIR MICROMATIC

The Micromatic was an updated version of the 1964 Micro-6 radio. Sold in kit form for the price of 50 shillings (£2.50/$4.65) and fully built for 60 shillings (£3.00/$5.60), it was to be Sinclair's last miniature radio. The Micromatic, like Sinclair's other radios, was really a very simple circuit based on easily obtainable and cheap components. Sinclair's "trick" for selling such an uncomplicated device was in the marketing and packaging. These little radios were very fragile and many must have been thrown away. Mint examples are very rare. Those still in kit form are even rarer.

1967
SONY ICR-100 RADIO

Sony's ICR-100 is recognized as the world's first integrated circuit radio. Smaller than a packet of cigarettes, this tiny device weighed about three ounces. It is easy to take such a small thing for granted today, but back in the late 1960s, the achievement was nothing short of remarkable. The radio was small enough to be carried on a keychain. Its case was a smart metallic design, years ahead of its time. It also used rechargeable batteries, but the charging units often went missing. Boxed examples are very rare.

1968
SINCLAIR NEOTERIC 60 AMPLIFIER

Sinclair Radionics had already ventured into the home-build hi-fi market by the time it launched the "space-age" Neoteric 60 stereo amplifier. This was the first Sinclair product to be sold in the shops, as well as through mail order. Its stunning design was created by Iain Sinclair, Clive's brother. It featured identical brass buttons and knobs, which gave a pleasing, uniform look to the front of the rosewood case. Unfortunately, there were problems with production, causing the amplifier to be dropped from the Sinclair range, but even today it is a beautiful device to own.

£50–£100 / US$95–$185

£50–£130 / US$95–$240

1968

SINCLAIR SERIES 2000 AMPLIFIER

The Sinclair Series 2000 range of products was the Cambridge-based company's second attempt to break into the retail hi-fi market. As well as the amplifier in this picture, the range included an FM tuner, a stereo decoder and speakers. The modern design was the work of Iain Sinclair, who was involved in many other Sinclair projects. The case of the amplifier was made from aluminium, which gave it a look far in advance of its competitors. The original cost was about £30 (US$55), which was very expensive. The 2000 range of products suffered from quality issues and were not a great success, but Sinclair continued to experiment with hi-fi and would later launch a range of quadraphonic products.

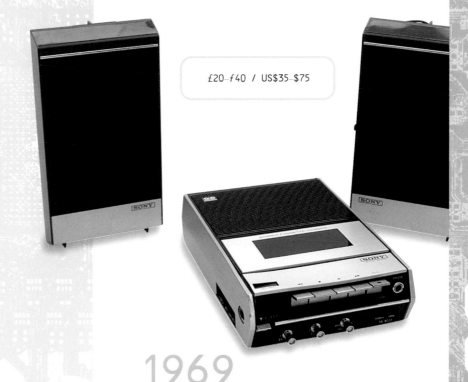

£20–£40 / US$35–$75

1969

SONY TC-124CS

The Compact Cassette, invented by Philips in 1963, was a very popular way of playing music on the move. The little cassettes could also be used for home recording, which made them doubly useful. By the late 1960s, many manufacturers were making cassette players, but high-quality portable systems capable of stereo reproduction were not common. Sony produced this portable stereo cassette system in 1969. The player, speakers, microphones and cables came in their own briefcase-sized carrying case. The player could be used without the external speakers, but they were needed for true stereo sound.

£20–£40 / US$35–$75

1969
SINCLAIR STEREO SIXTY

Sinclair's Stereo Sixty was a pre-amplifier system designed for use with other Sinclair hi-fi components. It was a popular system and was produced until 1975. The Stereo Sixty was often built into record decks, such as the Garrard example in the photograph. A pair of Sinclair amplifier circuits would have been included to complete the system. With a selling price of just under £10 (US$20), the pre-amplifier was an expensive purchase for the average hobbyist, and the components used in the circuitry were prone to the ingress of dust, which caused the controls to become crackly.

£20–£40 / US$35–$75

1970
JVC VIDEOSPHERE TELEVISION

Space travel continued to influence designers well into the 1970s. Space helmets were the theme for many products, but none was as obvious as this JVC Videosphere television. The 12.8 x 11.8 inch case could be mounted on a table or hung from the ceiling by a chain. Its black and white screen was hidden behind a visor, which completed the helmet-like look. The case came in a number of colours, including the very popular orange that seemed to cover everything in the 1970s.

1969
PHILIPS GF303 RECORD PLAYER

Rounded shapes and bright colours were themes of the late 1960s and early 1970s. The Philips GF303 portable record player illustrates both, with its circular case and a transparent lid produced in blue and red. Designed by Patrice Dupont, this little record player was just 11.8 inches in diameter and 4.7 inches high. Playback was monophonic, produced through a small speaker that was housed in a circular section in the centre of the lid.

£30–£70 / US$55–$130

£150–£250 / US$280–$465

1970

STAX SR–X EAR SPEAKERS

Japanese firm Stax, well known for its high-quality microphones, introduced these stylish SR-X "Ear Speakers" in 1970. They were based on an electrostatic design that used pairs of electrically charged grids to move a diaphragm, which produced the sound. You needed to use them with a Stax SRD-7 adaptor, to provide power for the headphones. The Ear Speakers rapidly gained legendary status in the audiophile community for their high-quality and transparent reproduction. Good examples are in demand today and, with spares readily available, they make an excellent and usable choice for vintage hi-fi collectors.

£100–£200 / US$185–$370

1970

HEATHKIT GR-98 AIRBAND RECEIVER

Heathkit was a major supplier of electronic kits, dating from the 1940s. This GR-98 Airband Receiver is just one example of the many thousands of kits made by the Michigan-based company. Heathkit's success was partly a result of the excellent manuals supplied with its kits. They avoided jargon and employed good diagrams to explain kit construction. Heathkit used the best-quality components and parts to give the builder every chance of succeeding with a project. The GR-98 receiver, one of Heathkit's smaller products, was an excellent radio when built well and even had a geared frequency knob for fine tuning.

£30–£40 / US$55–$75

1971

MAGNAVOX 8-TRACK RECEIVER

The 8-track cartridge tape system originated in the 1960s and, during the early 1970s, it was probably the leading format for in-car entertainment systems, until the Compact Cassette took over. In their day, 8-track systems, at home and in cars, provided high-quality reproduction, which was better than any other source. This Magnavox 8-track receiver would have been used in the home and dates from about 1971. An 8-track cartridge can be seen slotted in on the left-hand side of the panel. Some systems allowed recording on to blank cartridges. The 8-track system's popularity declined during the latter part of the 1970s, and disappeared entirely some time during the early 1980s.

£10–£30 / US$20–$55

1972
GRUNDIG SATELLIT 1000

This 1972 Grundig Satellit 1000 short-wave radio was an iconic example of the German company's craftsmanship. It ran on batteries or mains power and, while it was heavy at about 15 pounds, it could be transported anywhere. Sound reproduction was excellent, partly due to the use of two speakers fitted in the case. A serious radio for professional users, the 17-waveband Satellit 1000 provided connections for external aerials and had an aerial trimming control. If there was a radio signal in the air, the Grundig Satellit 1000 could tune into it.

1972
BANG & OLUFSEN BEOGRAM 4000

Danish company Bang & Olufsen produced the most unique designs of audio equipment ever seen. As well as their beautiful looks, B&O products were also technically advanced. The Beogram 4000 record turntable is one of its most famous products, in terms of technical wizardry. It used a linear-tracking tone arm, which, rather than moving across the rotating vinyl in a radial fashion, tracked across the record in a perfectly straight line. Everything on the device was automatic: all you had to do was place the record on the platter.

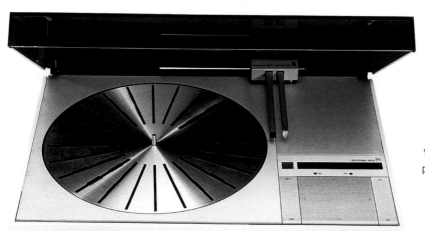

1972
POLAROID SX70 LAND CAMERA

The Polaroid SX70 was a folding single lens reflex camera. The entire device folded magically into itself to become as thin as its base. As well as a remarkable piece of engineering, the SX70 was a work of art. It was the first Polaroid camera to use motors to eject exposed film and the first to use film that didn't require peeling. Early models had manual focusing, but later designs used advanced ultrasonic technology to ensure a sharp picture. Nothing like it has been made since and the camera is still in demand by collectors.

£40–£120 / US$75–$225

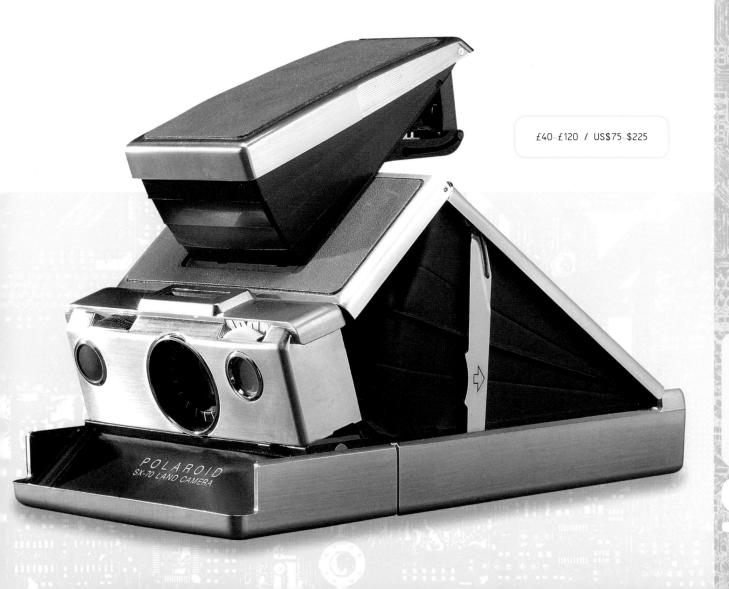

1972
PULSAR TIME COMPUTER

The first commercially available LED watch was marketed in 1972 by Pulsar, a division of Hamilton Watch. The company placed a full-page ad for the timepiece in *The Wall Street Journal*: its price was US$2,100 (£1,140). The case was available in stainless steel, 14k plated gold or solid 14k gold. The first examples were sold at Tiffany's in New York. The watches were beautifully made and many working examples survive today. The one shown here has a solid gold case and is one of the rarest examples so commands the highest price. Beware of lesser versions: LED watches became very common in the 1970s and many were cheap and of poor quality.

£300–£1,500 / US$560–$2,785

1972
LINN SONDEK LP12

£250–£1,400 / US$465–$2,585

Linn's Sondek LP12 record turntable has been in production for more than 30 years, such is the beauty and high quality of the device. Linn was one of the first makers to suggest that the music source, the turntable, should be the primary consideration when building a component audio system. The belt-driven turntable had a stainless steel chassis set into a solid wooden plinth. The heavy platter provided high mass, aiding speed stability. Although it is possible to spend huge amounts on turntables today, the modestly priced Linn Sondek is still the first choice of many audiophiles. The finish of the case, type of power supply, the tone arm and cartridge type can affect the value greatly.

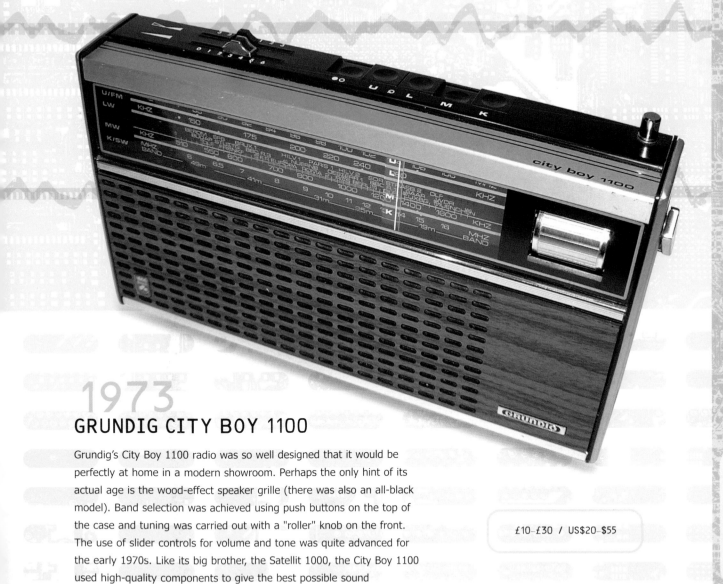

1973
GRUNDIG CITY BOY 1100

Grundig's City Boy 1100 radio was so well designed that it would be perfectly at home in a modern showroom. Perhaps the only hint of its actual age is the wood-effect speaker grille (there was also an all-black model). Band selection was achieved using push buttons on the top of the case and tuning was carried out with a "roller" knob on the front. The use of slider controls for volume and tone was quite advanced for the early 1970s. Like its big brother, the Satellit 1000, the City Boy 1100 used high-quality components to give the best possible sound reproduction.

£10–£30 / US$20–$55

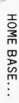

HOME BASE...

81

1974
PHILIPS N1502 VIDEO RECORDER

This Philips video recorder originally retailed at about £700, the equivalent today of around £4,000 (US$7,390)! It would have been a rare sight in a private house and was more likely to have been seen in a university or office. The proprietary Philips cassette slotted into a tilting lid, seen on the top left of the player. The electronics were advanced and used an LED clock with a battery back-up to cater for power blackouts, which were common at the time. Up to 70 minutes of broadcast could be recorded if you used the longest tape available.

£40–£60 / US$75–$110

1975
NAIM NAP250

The number of knobs, controls and lights on a music system was often seen as an indicator of cost and quality: the more of everything, the better. Naim ignored this trend completely when it designed the NAP250 power amplifier. The only requirement was to be able to switch the unit on and off, hence a single power button. The 17 x 11.8 x 3.1 inches unit was unconventional and clearly meant to be heard, rather than seen. It has been so successful that Naim is still selling the model, albeit with an updated case design.

£500–£2,300 / US$930–$4,245

1975
OMEGA TIME COMPUTER LED WATCH

The 1970s was a stressful time for established timepiece manufacturers. The success of LED digital watches was such that makers like Omega, famous for high-quality mechanical timepieces, launched electronic products to protect market share. This Omega Time Computer used a movement common to Pulsar's LED watches and was available in various metal cases, including solid gold, of which a limited number were made. The craftsmanship was superb and, like original Pulsars, the Omega would have been a very expensive item to buy.

£400–£2,500 / US$740–$4,650

1975
SINCLAIR BLACK WATCH

Clive Sinclair launched his Black Watch in 1975 as a kit, priced at £17.95 (US$33). It was also sold fully built for £24.95 (US$45), an option many hobbyists would come to wish they had taken up. The kit was notoriously difficult to construct and many non-working watches were sent back to Sinclair for replacement. There were other problems: accuracy was prone to temperature variations; the electronics were affected by static electricity from clothes; battery life was dismal; and the case would come apart all too easily. The Black Watch had a devastating effect on the company, but Sinclair continued to trade and never produced another LED watch. The Black Watch is so sought after that even non-working examples command decent prices.

£100–£400 / US$185–$740

1976

£40–£80 / US$75–$150

BANG & OLUFSEN BEOMASTER 1900

The stunning look of this Bang & Olufsen was the work of Jacob Jensen, the man behind many iconic B&O designs. The Beomaster 1900 was an FM receiver/amplifier and looked like nothing else on the market then, or since. There were no protruding knobs or switches; instead, touch-sensitive controls were sited along the front edge. Under the lid, however, there were normal knobs and switches, kept out of sight so as not to spoil the clean, space-age looks. The award-winning design was displayed at the Museum of Modern Art in New York and set B&O's look for many years after.

1976
GIRARD PERREGAUX CASQUETTE

The LED watch revolution had such a profound effect that even the most respected of mechanical timepiece manufacturers dipped their toes into the new electronic waters. Girard Perregaux made this stunning "Casquette" digital watch in around 1976. It was available in gold, stainless steel or, as shown in the photo, with a case made from ultra-modern glass-fibre-strengthened polymers. Note the LED display's orientation, which is viewed from the side of the case. It was designed so that it could be seen while driving. It would have been an incredibly expensive item when new.

£600–£1,000 / US$1,110–$1,855

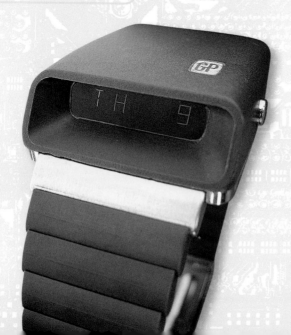

£50–£300 / US$95–$560

1977
SEIKO LCD WATCH

Seiko launched one of the world's first Liquid Crystal Digital watches in 1972. By 1977, the company was one of the leading sellers of LCD timepieces and offered a large range of models. LCD watches had a much better battery life than LED models. They were also more reliable and able to offer many more functions. Popular features at the time were stopwatch functions, the ability to tell the time in different time zones (like the model shown here) and even melodious alarms that played your favourite tune. They were hugely popular and, for a time, it looked as though the ordinary analogue watch would disappear under the digital onslaught.

1977

SINCLAIR MICROVISION TELEVISION

Clive Sinclair had a prototype pocket television as far back as 1966, but it took a decade for his dream to find its way into the marketplace. A massive sum was invested in the first pocket television, the Microvision, and as a result the sale price was high, starting at £250 (US$465). It used a six-inch-long cathode ray tube and a two-inch screen. The complex electronics were squeezed into a case just 3.9 inches wide, 5.9 inches deep and 1.6 inches high. The case also housed the rechargeable battery. As well as being one of the world's smallest televisions, the Microvision was also the first ever multi-standard receiver, working on both VHF and UHF bands. It was sold complete with various power supplies, which functioned in most countries, a case and a sun visor.

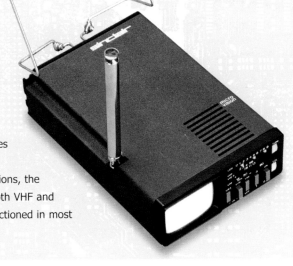

1979

SONY WALKMAN

The first Sony Walkman personal cassette player appeared in 1979, the result of the amazing foresight of Sony's Masaru Ibuka and Akio Morita. The first model, the Sony TPS-L2, was a massive success worldwide and was in such great demand that customers would pay a premium to obtain one. Within two years 1.16 million units were sold. New Walkman models were launched frequently to maintain the public's interest. Nowadays, practically everyone owns, or has owned a personal music device inspired by Sony's first foray into "personal hi-fi". Good examples of the first models are hard to find as they were generally used until wear and tear retired them.

£30–£100 / US$55–$190

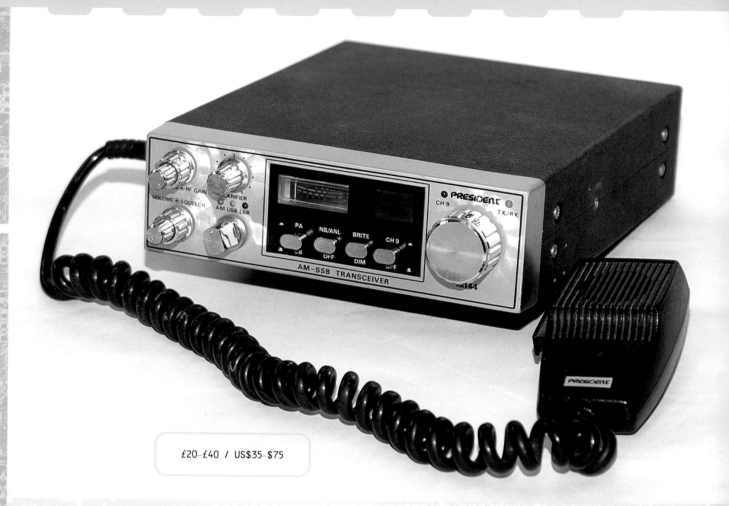

£20–£40 / US$35–$75

1979
PRESIDENT CB RADIO

Before the days of the cell phone, the only way to communicate over the airwaves was by two-way radio. The Citizens' Band frequencies allowed the man in the street to use radio equipment with little prior experience. CB radio became very popular in the 1970s, to such an extent that Hollywood made some well-known CB-inspired films. This President CB radio is a typical example of the type of set available at the time. A large knob controlled the channel, which was displayed on a large, bright LED display.

1981
CITIZEN LCD WATCH

Citizen, one of the world's largest manufacturers of watches, was a leading exponent of LCD watches in the 1980s. This unusual model used LCD elements to emulate an analogue watch display. During the early 1980s, the analogue watch started to fight back against the digital invaders and this LCD example shows that the fashion was swaying back to watch hands. The Citizen had the usual refinements of an LCD watch: an alarm, a stopwatch, two time zones and a back light for telling the time in the dark.

1982
PHILIPS VLP700 LASERVISION PLAYER

Philips's LaserVision format used 11.8 inch discs that looked like enormous CDs. These discs could hold up to 55 minutes of programming on each side, so while many feature films fitted on to a single disc, many others needed two. The LaserVision system used an analogue method of recording data, unlike modern CDs, which are digital. The model shown, a Philips VLP-700, used a large helium-neon laser tube. The casing carried many warnings about the danger of laser light. The system was not a major success, partly because the original selling price was about £550 (US$1,020).

CD100 COMPACT DISC PLAYER

PHILIPS

PUSH TO OPEN

ON/OFF

PROGR 1 2 3 4 5 6 7 8 9 10 11 12 13 14 15 ERROR

TRACK

PAUSE TRACK PROGRAM SELECTOR REPEAT

PLAY ▶ PAUSE STOP SELECT STORE ◀◀REV FWD▶▶

CANCEL REPEAT

NEXT CM SEARCH

1982

£30–£60 / US$55–$110

PHILIPS CD100 CD PLAYER

The Compact Disc has been with us for so many years that many people today simply won't remember vinyl LPs. It is hard to believe now, but CDs had a bit of a struggle to become the media format of choice. They used tiny pits in the disc surface to encode music digitally. This "sampled" delivery of music was frowned upon by many audiophiles of the time. This Philips CD 100 player was one of the first available. In those days, the preference was for top-loading designs, but these soon changed to the front-loading players we are familiar with today. Note the simple controls on the Philips machine: track searching was simple and similar to the tape decks of the time.

1982
SONY WATCHMAN POCKET TV

Sony's Watchman pocket television was a beautifully designed piece of equipment. The use of a clever "squashed" CRT allowed it to be housed in a compact case. The screen was black and white, and capable of a very clear picture. Tuning was carried out with a knob built into the top corner of the case, with an aerial located on the opposite side. A number of Watchman styles were made, but this use of CRT technology was soon to be superseded by LCD screens, which Sony adopted for its later pocket television models.

£15–£100 / US$30–$185

1983
SWATCH WATCH

The Swiss watch industry was under pressure from electronic watches made in the Far East. The mechanically dominant Swiss firms needed a strategy to reclaim their superior position as watchmakers. The result was a consortium of Swiss watch companies, who launched the Swatch. The concept was brilliant. A plastic watch with just 51 parts that, because it was sealed and therefore not repairable, was sold as a throwaway fashion accessory. Models were released in various colours and designs, with frequent limited editions feeding collectors' interest. Swatch would later launch new ranges and also branch out into other products and designs. The value of the rarest models can be very high.

£35–£55 / US$65–$100

1984

BRAUN VOICE CONTROL ALARM CLOCK

Braun has long had a reputation for excellent design and quality. The German company used famous designers, such as Dieter Rams, to produce an eclectic range of products, from the humble electric shaver through to sophisticated home entertainment systems. These were carefully designed with form and function in mind. Braun's 1980s Voice Control Alarm Clock is a good example of this approach. Rather than having to fumble for the clock when the alarm is activated, the sleepy owner had merely to shout to put it into snooze mode (note the microphone at the 10 o'clock position). The green bar at the top of the clock would eventually have to be pushed to silence the clock completely.

1984

SINCLAIR FLAT SCREEN TELEVISION

Sinclair's final television was the aptly named FTV (Flat Screen Television). The FTV used a flat CRT display, which was activated by electrons fired from the side and on to the front of the screen. The flat screen had to be viewed through a fresnel lens, which formed the screen, to restore the final picture to the correct dimensions. Power came from an expensive flat battery pack that slotted into the rear of the case. The original selling price was just under £80 (US$150), cheaper than the competitors at the time. The FTV worked well but was not a commercial success.

1985
SINCLAIR C5

£400–£650 / US$740–$1,210

The 1985 Sinclair C5 was meant to be a revolutionary new form of transport. Even though it worked, it was ridiculed by the press, who made jokes about it being made from "washing machine parts", a myth attributable to the vehicle being built for Sinclair by Hoover Ltd, the well-known vacuum cleaner company. A huge amount of research and development went into the C5. It was made from modern materials, could travel under its own power and was tremendous fun to drive. However, riding the C5 on a main road was a scary experience since the driver's head-height was very low and visibility was an issue for other road users, despite the tall flag that could be used. The C5 was a flop and was to be forever linked as such with the Sinclair name.

1986
CITIZEN LCD POCKET TELEVISION

£20–£40 / US$35–$75

This cigarette-packet-sized television was made by Japanese firm Citizen. It was one of the first televisions to use LCD for the display mechanism. You opened the lid to view the screen, which used ambient light to produce an image through an LCD panel, reflected on a mirror inside the case. The picture was black and white, and of surprisingly high quality for such a small device. The low power consumption of the LCD screen meant that battery life was very good. Owners could purchase a back-light for viewing at night.

Liquid Crystal Color Television
TV-400

COLOR

CASIO

CASIO

TV-400
POCKET COLOR TELEVISION

Simulated
TV picture

NTSC American
standard (M-VHF) s

£10–£25 / US$20–$45

1988
CASIO TV-400 COLOUR LCD POCKET TELEVISION

The Casio TV-400 was one of the first true pocket-sized colour televisions. The two-inch screen used a colour LCD display, which was illuminated from behind to give a bright picture in most conditions. Tuning was automatic, controlled by two buttons at the front of the case. The TV ran on four AA batteries, which would last for a few hours before they needed to be replaced. Extended viewing required the use of an optional power adapter. The 5.1 x 3.1 x 1.25 inch case was very small and the TV-400 can still be considered one of the smallest colour televisions ever made. Models varied according to the broadcast system of the country in which they were sold.

1988

SEIKO SPEAKING WATCH

This rare example of a Seiko speaking watch was made in around 1988. Most of the watch's case was given over to the speaker grille, leaving only a very small LCD display beneath it to show the time. Alongside the LCD screen was the "speak" button: you pushed it and a synthetic female voice told you the time. An advanced feature allowed owners to select different languages. One wonders whom the watch was targeted at – the technophile watch lover or perhaps a person with restricted sight. The watch, no larger or thicker than any other digital watch, also had an alarm and chime feature.

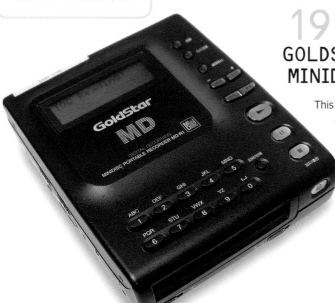

1992

GOLDSTAR MD-R1/SONY MZ-1 MINIDISC

This Goldstar personal Minidisc (MD) player/recorder was also sold as the Sony MZ-1. Together, they were the first MD players on the market. Like all new technology, they were expensive items, selling for about $700 (£380). These early MD players were quite big compared to modern versions, which can now be just millimetres larger than the discs themselves. The Goldstar had a 10-second shock memory feature, bass boost and the ability to label tracks with up to 21 characters. At 4.5 x 5.5 x 1.7 inches, the unit was just about pocket-sized and included an optical input/output for connection to a home-based music system.

£30–£60 / US$55–$110

1994
SWATCH CELLULAR

Swatch had enjoyed spectacular success with its innovative range of watches. In 1994, just over 10 years after its first watch was introduced, Swatch produced this cellular telephone. Described by the Swiss firm as "neat, light and easy to use", the colourful design was a refreshing departure from the competition, which was invariably drab. The phone was available in three colours: Blue Marine, Green Waves and Pret-A-Porter (the red one). Mobile phones are among the fastest-moving technology products and of great potential for forward-thinking collectors. Designer items like the Swatch Cellular are destined for future stardom in the collecting world.

1995
BAYGEN FREEPLAY WIND-UP RADIO

£20–£50 / US$35–$95

Trevor Baylis, the inventor of the Baygen Freeplay Radio, came up with the idea while watching a documentary on South African health problems. He realized that information flow regarding AIDS amongst poor communities would improve if radios didn't have to rely on batteries, and could be used more cheaply. The clockwork radio was the result. It used a spring-powered generator, good for more than 6,000 hours of use. The mainspring was replaceable, giving the eco-friendly receiver an indefinite life. You simply wound the handle for up to 30 minutes of listening. Despite the brilliance of the idea, it took Baylis many years to bring the radio to market, but when it arrived, it made him one of the world's most famous inventors.

1995
MUSICAL FIDELITY X-RAY CD PLAYER

£450–£550 / US$835–$1,020

This Musical Fidelity X-Ray CD Player had an interesting design, which was very different from any other player available then, before or since. It was a high-end audiophile product with complex 24-bit digital-to-analogue circuitry. This was meant to deliver the best reproduction possible from the Compact Disc format, which, being a sampled digital source, was seen as limited by many discerning listeners. The high-quality X-Ray would have had a sky-high price and, as such, would have partnered similarly expensive amplifiers and speakers.

1997

£30–£50 / US$55–$95

LEXON TYKHO RADIO

The Lexon Tykho Radio's looks – the work of French designer Marc Berthier – are unique and timeless. The case, which was completely covered in rubber, gives the award-winning AM/FM radio a strange, tactile feel. This selection of materials also meant that the radio had a degree of protection from water splashes. The Tykho came in a number of colours (but always with a contrasting grey aerial) and matched other Lexon products, so owners could furnish their homes with coordinated gadgets. The 5.5 x 3.1 x 1.6 inch case was compact and bijou – just right for displaying on the mantelpiece.

1997
SWATCH CORDLESS TELEPHONE

Just a few years after its angular cellular phone, Swatch launched this curvaceous digital cordless home telephone. The design illustrates the move from the angular style of the early 1990s to the smooth, almost organic, appearance adopted by products in the later years of the decade. Other than its cool, rounded looks, the Swatch Cordless was really just like any other digital cordless phone, offering clear speech up to 220 yards from the base unit. The handset sat in a cradle that itself sat in a larger home-base unit, allowing various configurations to be used.

£20–£40 / US$35–$75

1997
KARLSSON NIXIE TUBE CLOCK

£100–£150 / US$185–$280

Designed for Karlsson by Peter Van der Jagt, this Nixie tube clock is a masterpiece of design over function. The 6.9-inch-long case supported six 1.6-inch Nixie tubes, essentially valves modified to display 10 shapes, in this case numbers. The clock had a modern quartz movement and was set using two toggle switches at the back of the case. The use of Nixie tubes was usually identified with 1950s and 1960s equipment. Employing them in a late 1990s device was a clever juxtaposition of technologies. Karlsson's clock found favour with designer fans and was produced in limited quantities because of the restricted supply of tubes.

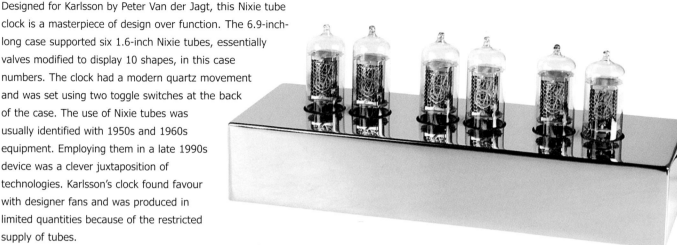

£10–£800 / US$20–$1,475

1999
EON CLASSIC LED TORCH

This space-age flat torch (flashlight) was designed by Iain Sinclair, who was responsible for many of the stunning designs in Clive Sinclair's product range. The Eon transcended conventional pocket light sources: solid-state, ultra-bright light emitting diodes were combined with special circuitry to produce a usable beam. The marketing claimed that "for average use you may never need to buy another torch". This credit card sized device was made in various materials, including solid gold and silver. As a true wallet-sized gadget with a designer pedigree, it is a sure-fire bet for collectors.

2000
BLUEROOM MINIPOD SPEAKERS

Blueroom, a spin-off of well-known speaker manufacturer B&W, brought us the Minipod range of speakers in 2000. Until then, speakers housed in anything other than an oblong box were the preserve of the wealthy. The Minipod changed that, with its brightly coloured, sometimes transparent cases that eschewed straight lines entirely. The shape wasn't just for looks – it had acoustic advantages, reducing unwanted reflections and vibrations. Initially released as a stereo pair of speakers, the Minipod range was later enhanced by the introduction of a subwoofer and a centre-speaker for home cinema use.

£180–£1,300 / US$335–$2,415

£40–£60 / US$75–$110

2001
PSION WAVEFINDER DAB RADIO

Psion, the firm famous for its range of palmtop personal computers, was behind this weird-looking DAB radio. DAB stands for Digital Audio Broadcasting and, if those behind it have their way, it will be the way most radios work in the future. The Wavefinder wasn't a radio in its own right. It was actually a computer peripheral that was connected to a PC by a USB cable. In conjunction with software, it received and decoded digital radio signals. It was fun to use, with the centre section of the unit displaying continually changing colours.

2002
LEICA DIGILUX 1 DIGITAL CAMERA

When the maker of the best film cameras in the world introduced a digital camera, the whole world watched. Leica's Digilux 1 was actually a combination of the German manufacturer's brilliant optics and electronics from Japanese firm Panasonic. The camera offered a four mega-pixel CCD and displayed results on a 2.5-inch LCD screen at the rear. One of the most important aspects of the camera was the full range of manual controls available, which offered the keen photographer total creative control. Among the millions of digital cameras made, only a few will become collectors' items in their own right. The Leica is one of them.

£400–£600 / US$740–$1,110

2002

£60–£110 / US$110–$205

SONY DSC U20 DIGITAL CAMERA

Sony was not being left behind in the digital camera race. In fact, the Japanese firm had some of the first digital cameras on the market during the 1990s. This DSC U20 was a tiny digital camera capable of good-quality photographs. The case was just 3.3 x 1.1 x 1.6 inches, but included a flash, Memory Stick slot and, most remarkably, a one-inch colour LCD display. In fact, the little camera included many of the functions found only in cameras many times its size.

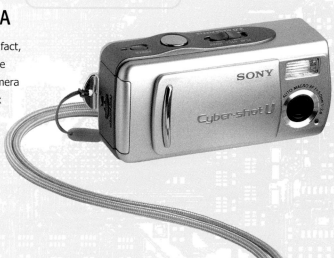

£70–£100 / US$130–$185

2002

PURE DIGITAL EVOKE-1 DAB RADIO

The Evoke-1 was one of the first commercially available DAB digital radios. When launched, it was so popular that demand outstripped supply and examples commanded a good premium on Internet auction sites. The radio was not just a technological marvel but was also brilliantly designed, making it an instant icon. It featured a digital display, illuminated by an alluring green backlight, with scrolling titles, artists' names and programme details provided by broadcasters. Although it was capable of stereo reproduction, the radio had just a single speaker. A second speaker, which could be plugged into the main unit, was an optional extra.

£100–£350 / US$185–$645

2002
APPLE IPOD DIGITAL MUSIC PLAYER

Even the most technophobic person will know about the Apple iPod, possibly the most iconic technological product that has ever existed. It was a superbly designed digital music player that contained a small hard drive, on which you could store digital music files. It was a massive success and its all-white design was to influence a new generation of products, not just in the digital music arena. New versions were released as technology improved, firstly with larger disk drives and then with new designs and features, such as touch-sensitive controls. Like the Sony Walkman, which announced the personal stereo to the world of the 1970s, the Apple iPod announced the arrival of personal digital players.

2003
CASIO EXILIM Z3 DIGITAL CAMERA

Digital cameras tended to fall into one of two camps: big, feature-rich semi-professional cameras or tiny, jewel-like fashion accessories. This Casio fell into the jewel-like camp, being tiny, shiny and pretty. However, it was a very capable camera, with a 3.2 megapixel CCD, integrated flash, a massive LCD screen and a sophisticated zoom lens, courtesy of Pentax. The zoom lens was particularly clever since it folded completely into the camera body by moving the optical elements out of the way. Here was a camera that looked as good as it was easy to use.

£100–£150 / US$185–$280

2004
NIXON THE DORK WATCH

At a glance, the Nixon The Dork could be taken for a 1970s watch, with its red digital display and large, angular case. The design was clearly meant to hark back to that decade, but the watch used the latest technology. The red display, despite looking like the old LED type, was actually LCD. The large button, when pushed, made the watch speak the time. The alarm used a number of voices, including that of famous skateboard star Tony Hawk. It was very popular with the surfing fraternity, which was ironic since the watch was not at all resistant to water. But it was still a great-looking watch, even if the battery lasted for only a few months.

£50–£70 / US$95–$130

PLAYTIME

The way we entertained ourselves underwent a dramatic change in the last quarter of the twentieth century. Before the arrival of affordable electronics, our desire to play was satisfied by cards, board games, musical instruments and sporting pursuits. But that was all about to change. The coming electronic revolution was to have a dramatic effect on our leisure time.

The first signs of technological influence on toys and games appeared in the 1960s. Computer programmers were using spare time on mainframe systems to code adventure games. Electronics was changing the way music was made. One of the best-known toys of the 1960s, the Stylophone, used a basic electronic tone generator to produce simple music. Transistors allowed small radios to be carried around – some could actually be worn – so you could listen to rock and roll wherever you happened to be.

The 1970s opened the technology floodgates, swamping us with toys and games based on the latest advances. Home video console systems were sold as early as 1972 but they were relatively expensive. It was the arrival of Atari's Pong game – and its subsequent 2600 console – that caught the public's imagination. They bought the new video games in their millions. Gaming arcades were also transformed. Out went the pinball machines, in favour of the new video games with names like Space Invaders, Galaxian and Pac Man. Traditional games, such as chess and checkers, were given the electronic treatment and had computer circuitry that played the part of the opponent. By the end of the 1970s, home computers began to appear, initially as hobbyist kits but quickly followed by commercially viable machines. The world had changed.

The 1980s was a decade of even greater change. Home computers, especially those made by Sinclair and Commodore, were to transform many youngsters into the IT specialists who are shaping our world today. Computers started to appear everywhere, but especially in the machines that would

entertain us. Every child was playing with electronic games. When they weren't playing on their home computers or consoles, they were transfixed by their handheld electronic games. Technology was moving fast.

With each passing year, an improvement in display technology made last year's equipment obsolete. New products appeared every month and the public bought them. Anything that had been around for a year or two was old technology and was either thrown away or put in the attic. Ordinary toys were being left behind. In a bid to catch up, manufacturers used any excuse to squeeze something "electronic" into a tired product, hoping it would sell on the back of the technology craze. By the end of the decade, practically every child had a game console of some sort, but the world was about to change again: the Game Boy had arrived.

They were heady days as we moved into the 1990s. The new generation of handheld cartridge-based games, with Nintendo's Game Boy taking the lead, brought about an overnight transformation in the way children played. The larger console-based systems were improving year on year, bringing new levels of complexity to games. Display technology continued to improve, which allowed handheld consoles to work with better graphics and colour. Quirky electronic toys appeared and disappeared in quick succession, sparking crazes on a scale not seen since the hula hoop. One of these was the Tamagotchi, an electronic pet housed in a plastic egg. At one point it seemed as though everyone had one, but then they disappeared as quickly as they had arrived. By the end of the 1990s, entertainment robots appeared. Some were very expensive marvels of technology directed at well-heeled adults, others cheap versions made to appeal to children.

As the new millennium dawned, manufacturers were building games into PDAs and cell phones. The first ten years of the twenty-first century are likely to bring as much change as the past thirty. Wherever technology takes game playing, a trail of machines highly desirable to collectors will undoubtedly be left behind.

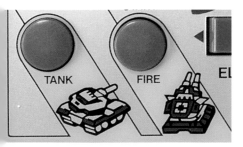 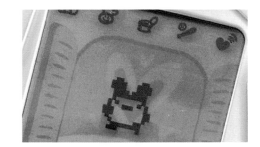 

£30–£60 / US$55–$110

1972
PANASONIC TOOT-A-LOOP

The 1970s – the decade of fun – demanded that products were bright and bizarrely shaped. Panasonic's contribution was the Toot-a-Loop, a radio in the form of a bangle that you could wear. It came in a variety of bright colours and was quite a clever design. It swivelled at its narrowest point to reveal the controls, which were hidden when the bangle was shut. There was no other radio quite like the Toot-A-Loop and it has since become very popular with radio collectors, who like to own the whole set of colours. Boxed examples fetch the highest prices, with red seemingly the most common colour.

1968
STYLOPHONE

It was small, plastic and made a strange noise, but the public went mad for it. The Stylophone was a massive hit in the UK in the late 1960s and early 1970s, not least because of the clever marketing tie-up with Australian star Rolf Harris, who promoted the musical device on television. In fact, it was voted Toy of the Year for the UK in 1970. The Stylophone was invented by Brian Jarvis, who worked for Dübreq, a company involved mainly in recording technology for film and broadcasting. The first examples were handmade and cost £9, the equivalent of about £100/US$185 today. Stylophones are always in demand and sell readily for good prices. Various models were made, including a large, professional-level one. The best examples to collect are boxed with the original booklet and record.

£40–£80 / US$75–$150

1972

MAGNAVOX ODYSSEY

The Magnavox Odyssey was the first video game system available to the home buyer. Launched in 1972, it was remarkable for being built solely from discrete electronic components. On the television screen the Odyssey's circuitry generated a couple of bars to act as bats and a spot that played the role of a ball. There were two ways of varying the game. Firstly, you could change "cartridges", which affected the appearance of the ball and bats on the screen; the cartridges were merely switches and contained no electronic components. Secondly, the Odyssey provided transparent coloured overlays (it did not have colour circuitry) that gave the simple screen more complexity for the games; the overlays came in a couple of sizes to cater for different television sets. Some of the games required props, which were supplied with the console and included money, counters and cards. The Odyssey sold for £55/$100 in 1972 and was made until 1974, by which time more sophisticated video consoles had arrived.

£70–£100 / US$130–$185

£10–£40 / US$20–$75

1975
ELECTRONIC CHESS

The first electronic chess product is usually attributed to Fidelity Electronics, circa 1975. The original inspiration for an electronic chess game is said to have come from Fidelity's owner after watching an episode of "Star Trek". Chess is a game that electronic logic was ideal for emulating. As more sophisticated circuitry was developed, and production costs fell, more companies started to produce electronic, or "computer", chess sets. The Mephisto model shown here is a later and more sophisticated example (note that it uses an LCD display, whereas the original Fidelity machines would have used LED displays). Some sets had sensors on the board so that the computer circuitry could keep abreast of each move. One expensive model from the 1980s even moved the pieces on behalf of the computer. When collecting chess games, it's important to ensure that the original pieces are complete, so it's best to buy systems in as new a condition as possible and with their original boxes.

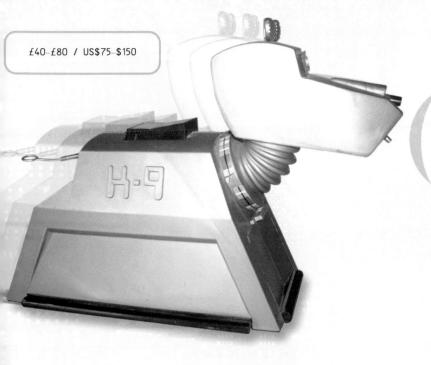

1976
PALITOY TALKING K-9

K-9 was the canine robot companion to Doctor Who, the time-travelling hero of the famous BBC TV science fiction series. K-9 was introduced to "Doctor Who" in the 1970s in an attempt to endear the television show to a younger audience. The robot dog used to offer the Doctor advice in a high-pitched robotic drawl. He could sense danger using his radar-like ears, which oscillated wildly in times of trouble. When necessary, K-9 could fire laser blasts from his nose-mounted weapon. Palitoy, the company behind Action Man and GI Joe, were quick to capitalize on K-9's popularity and marketed this talking model. Rather than using sophisticated, and therefore expensive, voice-synthesis circuitry, K-9's voice was produced using a tiny record within the model's body.

1977
MILTON BRADLEY LOGIC 5

While other toy manufacturers were still wondering what to do about the electronic revolution, Milton Bradley's responses were already on the shelf. One of their first electronic offerings was the quirky Logic 5 game (also marketed as the Comp IV or the curiously spelt Pythaugoras). Logic 5 looked like a child's calculator but it was actually a testing puzzle, so much so that it was advertised with the slogan "I am programmed to beat you". The object of the game was to guess the number that Logic 5 had in its memory. You typed in your guess and Logic 5 indicated which numbers were correct and in the correct place, and which were in the wrong place. The results were given as a simple row of LEDs. Logic 5 is one of the rarer Milton Bradley games today and very difficult to find with the original box.

1977
TELE-TENNIS

The arrival of Atari's Pong game raised the public's awareness of electronic gaming because it used the home television: it didn't take long for shops to be swamped with similar products. The first models were humble black-and-white offerings, with simple one- or two-player tennis games. Later models had colour-screen displays and more games but they were all based on the simple bat and ball concept. The most advanced machines allowed the use of a light-gun for simple target practice and shooting games. These consoles were sold in huge numbers so collectors must concentrate on well-known makes and go for pristine boxed examples.

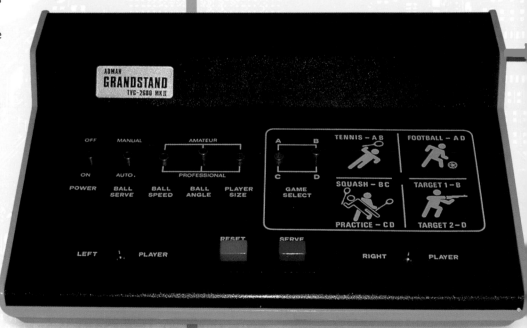

£10–£30 / US$20–$55

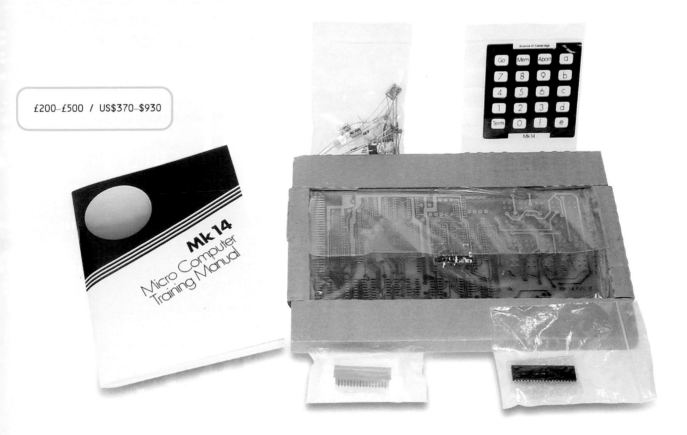

£200–£500 / US$370–$930

1977
SCIENCE OF CAMBRIDGE MK14

The Science of Cambridge MK14 (the MK stood for Microcomputer Kit) was a computer of sorts. It was more of an educational kit to be played with by those curious about computing. Despite its limited capabilities, the MK14 was one of the most important computers of the era. Its success spurred Clive Sinclair to pursue the development of his range of computers, which had a massive impact on the home market. The £40/US$75 MK14 kit comprised a printed circuit board, an LED display, a keypad, instructions and the electronic components. Despite being based on a calculator chip, the MK14 allowed people to grasp the basics of low-level computer programming. About 50,000 kits were sold and it is highly probable that most were assembled, so unbuilt kits are a very rare and valuable to collectors.

£40–£100 / US$75–$185

1977
ATARI 2600

The Atari 2600 was a cartridge-based video game machine for use on a home television. It came with two joysticks and a pair of paddles. Atari initially had the monopoly on game production – there were nine games during the first year of production – but later allowed third-party developers to market games. This meant that many new software companies gained a foothold in the games market; some are still with us today. Tens of millions of 2600 consoles were sold and millions have survived to feed the retro-gaming market today. The original wood-effect console, such as the one pictured, must not be confused with the later silver version, which is also called the 2600 but not so desirable to collectors.

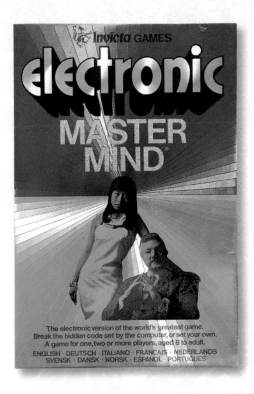

1977

ELECTRONIC MASTER MIND

In the latter part of the 1970s, there was a frenzy of activity to capture a slice of the new futuristic market with something "electronic". Toy manufacturers were in danger of being left behind and many took the plunge, launching electronic versions of established toys and games. Invicta, who already had a massive success with its Master Mind puzzle, decided to launch an electronic version. Electronic Master Mind had the appearance of a pocket calculator, which in essence it was, albeit with different programming. Instead of using colours, like the original game, the electronic version used numbers that were displayed on the LED screen. It was a well-built product and came with a wallet, paper and pen.

£10–£20 / US$20–$35

1977

MATTEL ELECTRONICS FOOTBALL

Mattel Electronics Football was the company's second electronic handheld game. Discrete LEDs were used to represent a football field on a small screen. Above the field screen, an array of calculator-like digital displays conveyed down, yards to go and field position information. The game was actually very complicated and capable of realistic football strategies. Power consumption meant that a relatively expensive 9-volt battery had to be used. Sales of the game were initially slow but soared in 1978 as the electronic games market took off. A huge number were made but, as with many toys, a huge number must also have been thrown away, with the result that Football and Mattel's other handheld consoles are attractive to collectors today.

£15–£25 / US$30–$45

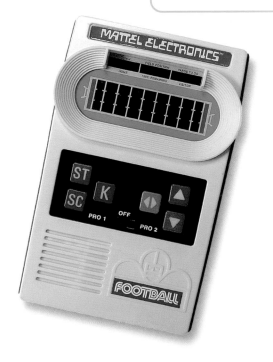

1977
TOMY BLIP

Before the use of electronics for synthesizing motion in a toy, manufacturers employed mechanical solutions. Tomy was very experienced with mechanical mechanisms and had used them in its successful Pocketeer range of games, among others. Its 1977 Blip game was a peculiar marriage of mechanics and electronics, in that gameplay was produced using a clockwork engine but the display used a single LED to represent the ball. Blip was not alone in using this combination of technologies and is a typical example of the enormous transition the toy world was undergoing in 1977.

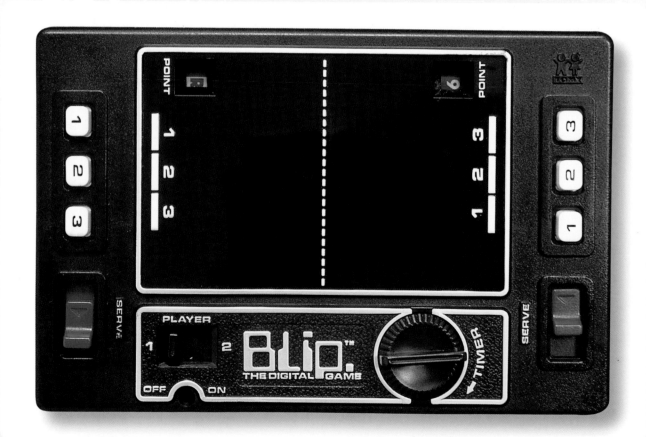

£10–£50 / US$20–$95

1978
MILTON BRADLEY SIMON

Designed by American firm Marvin Glass & Associates for Milton Bradley, the Simon game is perhaps the most recognizable electronic toy from the 1970s. The basis for Simon was simple. Just four buttons lit, one by one, while sounding a tone; the player had to memorize the sequence and repeat it. As the sequence became longer, the game sped up. Shops could not get enough Simons to meet demand. Such was the game's popularity that clones were spawned and it is still possible to buy similar electronic games today. Milton Bradley also made a pocket version but it is the full-sized original game that collectors want most. Boxed examples are quite scarce and command the best prices.

1978
TEXAS INSTRUMENTS EDUCATIONAL CONSOLES

£10–£20 / US$20–$35

In 1978 Texas Instruments launched a range of educational toys that employed quite advanced voice synthesis technology. Its best-known product was called the TI Speak & Spell (top left in this photograph). The Speak & Spell was programmed with spelling games and could communicate using a friendly, synthesized, male voice. An LED display also showed words and results. The earliest models had raised buttons, but these were soon replaced by a continuous membrane, which protected the electronics from drink spillage. Texas Instruments continued to bring out new products, such as the Speak & Read, Speak & Math and toys for teaching music skills. The small calculator-like device at the bottom of the picture is the TI Little Professor. This did not have voice synthesis but was an enormously popular mathematical educational toy. The first version used a red LED display, with later models using an LCD screen, just like the one shown here.

1979
ATARI 400

Atari was already enjoying success with its 2600 video games console when, in 1979, it released its first home computer, the 400 (the "400" name was derived from the amount of memory, 4KB, originally planned for the machine). The 400 had an integrated cartridge slot for ROM-based games and a strong bias toward games with powerful sound and graphics capabilities. To reduce costs, Atari used a simple membrane keyboard on the 400; a more expensive version with a "proper" keyboard was also available. The 400 came with up to 16KB of RAM and could support the joysticks and paddles used on the Atari 2600 console. Production ceased in 1982 when new machines appeared on the market.

£30–£60 / US$55–$110

1979
PALITOY MERLIN

Merlin was Palitoy's attempt to cash in on the electronic game craze. The telephone-like Merlin had eleven LEDs integrated into the panel of buttons. Using a combination of flashing LEDs and beeping noises, as many as six games could be played. These included Echo, which was similar to the copycat style of game introduced by Milton Bradley's Simon; three-in-a-row, a tic-tac-toe type of game; and secret number, which was similar to the Invicta Master Mind game. Although a well-built toy, Merlin required six AA batteries, which made it heavy for small hands and expensive to use. It sold in reasonable numbers so is fairly easy to find today.

£10–£20 / US$20–$35

1979
SPACE INVADERS

Ask any adult on the street to name an arcade video game and the odds are they will say "Space Invaders". The impact of Space Invaders on the video gaming world cannot be underestimated. Home versions of the game caused the sales of consoles like the Atari 2600 to rocket. The idea was simple. Rows of menacing aliens slowly descended on you and dropped bombs. You had to defend yourself with a gun and could seek cover in just four buildings. Every now and then, a UFO flew overhead, offering you bonus points if you destroyed it. To make the game even more exciting, your shelters were slowly demolished as the invaders' speed of descent increased. The objective was to destroy the aliens before they reached and destroyed you. The game was an astronomical success and is still hugely popular today.

£200–£700 / US$370–$1,300

1980
BALLY FLASH GORDON PINBALL

Pinball machine manufacturers were under considerable pressure from arcade machines such as Space Invaders and Galaxian. They needed extra tricks to keep their machines fashionable. So in 1980 Bally introduced a number of new features on its famous Flash Gordon pinball machine, based on the film from the same year. It was one of the first machines to use speech and incorporated three voices from the film to deliver messages, such as "Ignite death rays – 15 seconds", that were linked to game activity. There were two ball levels, each with a bonus system, and the upper level had its own single flipper. It was a great game to play then, and with just 10,000 made it is an expensive item for collectors today. Prices vary greatly depending on condition and the uninitiated should consider purchasing a fully restored machine.

£500–£1,500 / US$930–$2,785

1980
BANDAI MISSILE INVADER

The arrival of the Space Invaders arcade game in 1979 had a profound and lasting effect on all aspects of computer games. The plethora of new handheld computer games in 1980 was probably the result of Space Invaders' huge popularity. One of the better early handhelds was Missile Invader from Japanese company Bandai. The game itself was a clone of the Space Invaders idea and used the ubiquitous LED to represent aliens, missiles and UFOs. Bandai was a prolific maker of handheld consoles in this era and it is thought that the company made more than 100 electronic games.

1980
MATTEL INTELLIVISION

Mattel, already a successful producer of electronic handheld games, needed a product to challenge the Atari 2600. That product was the Mattel Intellivision, although it was actually quite different from Atari's console. It had better graphics, better sound and was designed to be modified into a home computer with the addition of a keyboard. The controllers were hard-wired to the console and, when not in use, could rest in their own slots in the main unit. There was a programmable keypad on each controller that allowed the Intellivision to have more complex game control than was offered by other makers' machines. The console was supported until 1990, by which time more than three million had been sold.

1980
ENTEX DEFENDER

Arcade games were having a major impact on the electronic games world. The huge success of Space Invaders had alerted games manufacturers to the commercial possibilities of clone games. Space Invaders didn't have the monopoly, because there were other major arcade successes – and one of those was Defender. Defender was a much more complex game than Invaders and involved piloting a spaceship in a two-dimensional world and destroying alien ships while rescuing stranded humans. It was immense fun and game manufacturers rushed to capitalize on its success. Entex had the best version in its table-top Defender game. It used a fluorescent display to emulate the original version, with a magnifying screen to give it a realistic arcade feel.

£30–£40 / US$55–$75

1980
NINTENDO GAME & WATCH FIRE

Nintendo was formed in the nineteenth century and initially manufactured trading cards. In 1975 the company co-operated with Mitsubishi Electrical to investigate the use of electronics for domestic games. Nintendo introduced its coin-operated video games in 1979 and its world-famous character, Donkey Kong, arrived in 1981. The Japanese company's dominance of the electronic gaming market moved into the home with the introduction of the exquisite range of LCD display devices called Game & Watch. The success of these would lead directly to the development of the Nintendo Game Boy, nine years later. Game & Watches were probably the first handheld games to use LCD displays and were built to a very high standard.

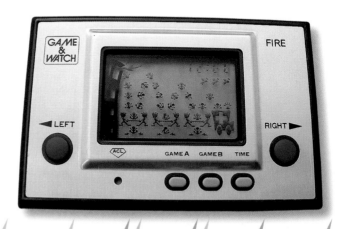

£20–£40 / US$35–$75

1980
SINCLAIR ZX80 AND ZX81

The Sinclair ZX80 is possibly the most famous home computer in the UK market. Sir Clive Sinclair had already experimented with a home computer, the MK14, but his ZX80 was a real computer with a keyboard (even if it was flat and tiny) and an output that could be viewed on a normal television. Little could be done with the ZX80 because it had such a small amount of memory, but it still sold well in both fully assembled and kit form. About 70,000 ZX80s were made, many fewer than the millions of its successor, the Sinclair ZX81. This fact makes the ZX80 a very rare machine and mint boxed examples regularly trade for large sums. Mint versions of the ZX81, meanwhile, attract much lower prices.

£150–£400 / US$280–$750

£30–£60 / US$55–$110

1980
BANDAI GALAXIAN

Space Invaders and Defender, both arcade classics, had spawned handheld clones. As the arcade industry blossomed, more games arrived and one of those was the classic Galaxian. Galaxian was a derivative of the Invaders idea but included more animation, better sound and better colour. It was arguably the greatest arcade game of its time. Bandai produced one of the best handheld Galaxian games, which not only offered involving play, but also presented itself in a high-quality futuristic case. It used a fluorescent display, which was becoming more popular in handhelds. Players had three minutes to defend their bases from hordes of descending aliens – nothing new but still an entertaining game.

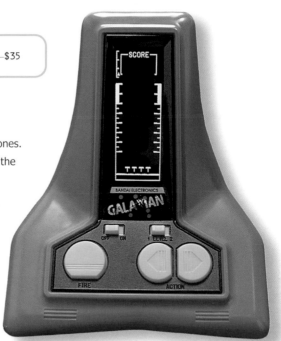

1981
COLECO PAC-MAN

Some game manufacturers built table-top versions of arcade classics using cases that looked like miniature versions of the real thing. The arrival of Pac-Man, thought by some to be the king of arcade games, inspired this Coleco table-top version. Note the style of the case, which is meant to give the player the feeling of being on a real arcade machine. The display used fluorescent elements, which were becoming standard on better quality machines. The original price of Coleco's Pac-Man was about £35 ($60), so it was an expensive item. That probably explains why it is so hard to find today.

GRANDSTAND ASTRO WARS

£15–£25 / US$30–$45

Astro Wars by Grandstand was actually licensed from Epoch, another well-known game manufacturer. Grandstand's game used the now-common fluorescent display for an Invaders-like shooting game. The screen – unusual because it was round – had a magnifying lens to give the game a big-screen feel. The quality of materials was noticeably poorer than those used on many of the more expensive games from the same era. It ran on four C-type batteries, with a socket for an optional power supply. Astro Wars must have sold in large numbers since it is a fairly common game in the collecting world today.

1982
CGL M5

£20–£50 / US$35–$95

The electronic gaming world shifted more of its attention to the home computer in 1982 and there was a marked increase in numbers available in the shops. The CGL M5 is one of the more peculiar machines from that year. It had a rubber keyboard, which was very similar to that used on the Sinclair Spectrum, but, unlike the Spectrum, it had a cartridge slot for games and other programs. The CPU was the extremely popular Zilog Z80a, running at 3.5MHz and serviced by 32KB of RAM. Unlike other home computers of the time, the CGL didn't have built-in BASIC, favouring cartridge-based ROMs to load programming languages.

1982
SINCLAIR ZX SPECTRUM

£30–£80 / US$55–$150

The most famous and most important home computer ever built, the Sinclair ZX Spectrum kicked off a computing legacy that still thrives today. Not even its creator could have realised the profound impact that this iconic machine would have. The Spectrum had a rubber keyboard and each key could perform as many as six different functions. Some people found this confusing, even annoying, but that didn't stop the Spectrum becoming the best-selling home computer of its time, if not ever. New small companies emerged to produce a huge array of accessories and peripherals for it. The Spectrum also spawned a fledgling software industry and led to the creation of many of today's best-known computer game manufacturers. The ZX Spectrum sits on top of the computing pyramid of fame.

1982
COMMODORE 64

£20–£50 / US$35–$95

The Commodore 64 was Abel to Spectrum's Cain. Never had the computing world been as divided as it was over the ZX Spectrum and Commodore 64. The rivalry lives on today, and the two groups of users are still arguing about which machine was better. Each was an 8-bit machine with colour graphic support. The Commodore had more memory (64KB) than the Spectrum (48KB), as well as a proper keyboard on which you could touch-type. Both had vast support from software writers and many games were produced, some for both machines. However, the Commodore held an advantage for those seeking more serious use. True, the Spectrum was capable of serious computing but the Commodore had the edge because of its grown-up design, sensible keyboard and better peripheral support.

DRAGON 32

£20–£50 / US$35–$95

Dragon Data, the Wales-based subsidiary of then-troubled toy manufacturer Mettoy, released the Dragon 32 home computer in 1982. The "32" name was derived from the 32KB of RAM within the machine, which, unusually for a home computer at the time, was based around a Motorola processor. The Dragon had a proper keyboard and some advanced features, so it was suitable for serious business use as well as playing computer games. A 64KB model was introduced later and sold for just under $400 (£215) in the USA. Dragon met with great success in the first few years, with some production moving to the USA. Despite this and other achievements, Dragon computers are not as well known as they should be.

1982

CAMPUTERS LYNX

£50–£140 / US$95–$260

One of the more obscure early home computers was the British-built Camputers Lynx. The Lynx was a fairly ordinary-looking machine, with a full keyboard housed in a plastic shell. It was a good computer and, unlike many home machines, could run the CP/M operating system. Circuitry was based around Zilog's Z80A processor with 48KB of RAM on tap. A later version came with 128KB of memory. However, the Lynx had no major distinguishing features and this may have been one reason for its lack of success. It was made for just a couple of years, which, combined with the small number sold, makes it a very rare item.

1982
MILTON BRADLEY VECTREX

The Vectrex stands alone in games console history as the only cartridge-based gaming system to include a built-in CRT monitor. The designer's intention was clear: to create as much real arcade machine feel as possible. The smart joystick unit could be stored below the screen when not in use and an optional second joystick could be plugged in for two-player games. The screen was black and white, but colour gaming was possible through coloured screens that slotted on to the front of the unit. The name Vectrex was derived from "Vector Graphics", the system used by the early Asteroids arcade classic. The original price of $199 (£110) made it quite an expensive purchase, although the price subsequently fell to about $100 (£55).

£50–£100 / US$95–$185

£20–£60 / US$45–$110

1982
NINTENDO GAME & WATCH DONKEY KONG JR.

Donkey Kong Jr. was one of the most popular of Nintendo's handheld LCD games. It was an arcade classic, up there with Pac-Man and Space Invaders, and translated very well to the handheld format. The huge popularity of Game & Watch kept the brand in production until the end of the 1980s, when only the arrival of the revolutionary Game Boy halted its production. By the mid-1980s there were versions that partnered games produced for the Nintendo Entertainment Console. This strengthened Nintendo's grasp on the computer games market. Values depend on condition and it is important to ensure that the serial number and battery cover are intact.

1983
NINTENDO GAME & WATCH POPEYE PANORAMA

Nintendo took its Game & Watch range to a new level with the introduction of the Panorama series. The Panorama was a larger game that used a clever folding mechanism and a mirror to reflect a colour image of the LCD screen. Six Panorama versions were made in 1983 and 1984, including the Popeye game shown here. Build quality was superb and with an original price of £20 (US$35), one of these would have made a very special present. The Panorama range included some of the rarer Game & Watches.

£50–£150 / US$95–$280

£10–£15 / US$20–$30

1983

CASIO MISSILE DEFENDER

Casio, the Japanese company better known for pocket calculators and watches, made a surprisingly large number of LCD games with a format similar to that of Nintendo's Game & Watch series. Despite this, the collector market for Casio games is tiny compared to the Nintendo Game & Watches market. The Missile Defender game shown here dates from about 1983 and was based on the arcade game of the same name. Casio was not so rigid as Nintendo with its game format. The button position, screen size, case size and other factors were all variable.

1983

GAKKEN TOM & JERRY

Gakken was another Japanese company that jumped on the electronic games bandwagon in the 1980s. It released a range of LCD games with designs that were similar to the Nintendo Game & Watch. Gakken branded some of these "Card Game" since they were fairly thin and easy to carry. However, Gakken's games failed to capture players' imagination as effectively as some of the competition's and lacked the big-name arcade links. Most collectors see Gakken's games as the poor cousin of those from other, more successful manufacturers.

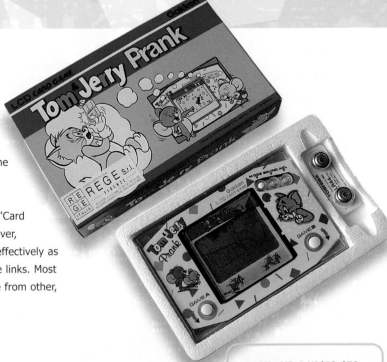

£10–£15 / US$20–$30

1983
TOMYTRONIC 3-D

Japanese company Tomy brought out this innovative 3-D LCD game under the name Tomytronic. Players viewed the screen through binocular-like lenses to see a 3-D arena that, depending on the version, could be a racetrack, a space war, a tank battle and so on. The image was produced by light passing through LCD panels (see the pane at the top of the case in the picture) and being reflected by mirrors into the player's view. The controls were easy to operate because they fell under the player's fingers as the game was held up to his or her eyes. Tomytronic 3-D games even had straps so owners could wear them like real binoculars.

1983
GAKKEN PINBALL

Pinball was another of Gakken's Card Game range. The fussy screen illustrates the subtle differences between the Gakken and designs created by Nintendo, which are much cleaner and easier on the eye. The game was perfectly playable and quite well made, but beside a Game & Watch, the difference in quality was very clear. Gakken's games, like Casio's versions, can be difficult to find compared to the Nintendo Game & Watch, but despite this, the values are significantly lower.

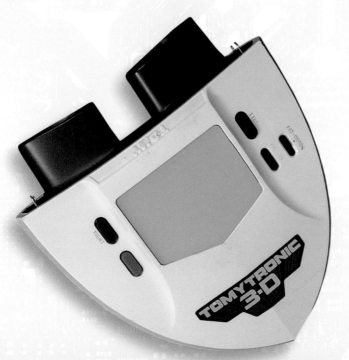

My home is Room 133 Bdg 93.
If you take me away then
write your name and date
on the white board.
Bring me back when I am idle.

NPL DNAC
461

1983

JUPITER ACE

The Jupiter Ace was built by some of the team responsible for the Sinclair ZX81. This may explain the striking resemblance to the Sinclair machine, although the Ace's keys are more akin to those of the Sinclair Spectrum. The Ace was unique because its resident programming language was Forth, a language associated with embedded systems and not seen as a hobbyists coding environment. The Ace was, therefore, clearly meant to be attractive to programmers rather than game players. This may be the reason that sales were poor. Nowadays, a Jupiter Ace is a rare sight, with examples snapped up quickly by eager collectors.

£150–£200 / US$280–$370

1983

GRANDSTAND FIREFOX F-7

£10–£20 / US$20–$35

The Firefox F-7 game used a fluorescent display with a magnifying screen to give players the feeling that they were flying through a trench on a space station, just like in the final battle scene from *Star Wars*. As with Grandstand's Astro Wars, Firefox F-7 was made under licence from Epoch, who also sold it, albeit with the less catchy names of Astro Thunder 7 and Galaga X-6. The game featured twin-speaker sound that, although not true stereo, did enhance the gameplay. There was also an off switch for the sound so that parents could be relieved from hours of repetitive space station storming.

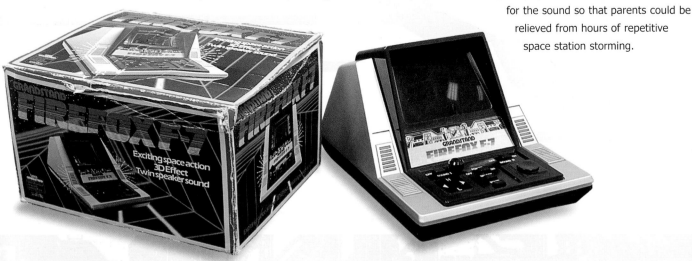

1984

ORIC ATMOS

£40–£70 / US$75–$130

The Atmos was the successor to Oric's earlier Oric-1. It was essentially the same machine but had a new case with a better keyboard and 48KB rather than 16KB of RAM. Like most other home computers, the Atmos was programmed using a version of the BASIC language. It supported eight colours, a 40 x 28 character display and sound. Although it was an attractive design, with a striking red and black case, there was nothing remarkable about it compared with any other machine of the time. Never a big seller, the Oric Atmos is a scarce machine today.

1984
AMSTRAD CPC-464

Amstrad, a company well known for selling
feature-rich products at economical prices, took
on the computing market with a range of
machines that included the CPC-464. The
"CPC" stood for Colour Personal Computer and
the 464 was one of the first computer packages to come
complete with storage and a monitor. It had its own version of BASIC
built in (a similar version appeared in the later PCW range of Amstrad
computers) and came with 64KB of RAM. Tape storage was built into
the main keyboard unit, but this was replaced in later models by a
much faster three-inch floppy drive. Strong marketing and a low price
ensured the CPC-464's commercial success and many were sold.

£30–£60 / US$55–$110

1984
CASIO MX-10

Casio's MX-10 home computer was one of a range
of MSX standardized machines produced by a
number of makers, including Sony, Toshiba and
Sharp. Had MSX standard machines
succeeded, we would all be using their
successors today, but they failed to attain
the level of success that they deserved.
The exact derivation and meaning of
"MSX" is the subject of much discussion and
probably worthy of a book on its own. Casio's MX-10 only just
met the strict standard and used calculator-type keys that
annoyed so many people on other machines. Not the best
MSX machine, but an interesting find for the collector.

£20–£40 / US$35–$75

1984
MASUDAYA ROBBY THE ROBOT

The Forbidden Planet, one of the greatest science fiction films ever made, introduced audiences to Robby the Robot in 1956. The metallic servant, which had more brains than the other characters in the film, was not the first robot to appear on screen – that honour fell to False Maria in Fritz Lang's *Metropolis* – but he was probably the first to fuel a toy robot industry that is still buoyant today. Japanese toy maker Masudaya released this Robby model in about 1984. The figure, which stood 16 inches high, used batteries to drive a voice system and lights. The Robby model spoke phrases from his film role, such as "...welcome to Altair 4, gentlemen..." and "...I am at your disposal with 187 other languages...". The voice was activated by pushing a button on the robot's chest panel. As Robby spoke, a light would blink in the grille that represented his mouth, just below the glass-domed head. Masudaya made other versions of Robby, some of them quite recently.

£40–£60 / US$75–$110

1984
SONY HB-101

Sony's HB-101 (HB stood for Hit Bit) was one of the famous Japanese company's MSX computers. The MSX standard was devised by a number of companies in an attempt to dominate the home computer market. Part of the standard was the directional cursor keys that you can see to the right of the keyboard. The Sony included a screw-in joystick that often became lost. This strikingly designed and futuristic computer had a pull-out carry handle for easy transportation and the case came in red or black, depending on the country of sale. Note the two cartridge slots at the top of the case.

1985
NINTENDO ENTERTAINMENT SYSTEM

Initially sold in Japan as the Famicom (Family Computer), Nintendo's first computer gaming console was released in Europe and the USA as the Nintendo Entertainment System, or NES. It was an 8-bit machine running on a 6502 processor, supporting up to 16 colours on the screen. The launch price varied between $200 and $250 (£110–£135), depending upon the package. The NES was a cartridge-based machine similar to other consoles of the time. The game cartridges, which were quite large, would slide into a holder that was hidden behind a hinged lid at the front of the machine. A huge number of games were made for the NES, some of which are regarded as classics. It was a fantastic console and is still great fun today.

£30–£50 / US$55–$95

1985
TOSHIBA HX-10

Toshiba was another Japanese company involved with
the MSX initiative and the HX-10 was one of its first MSX
machines. It had a fairly simple design compared to
other MSX computers. For example, the cursor keys to
the right of the keyboard didn't have the screw-in
joystick offered by some other MSX computers. Despite
any shortcomings, the HX-10 sold quite well, which
may have been because it was one of the first MSX
machines available outside Japan. MSX computers
were based around the Z80A processor running at
3.6MHz, with built-in MSX BASIC. MSX computers,
like the HX-10, were relatively expensive compared
to other machines of the time.

1985
CASIO PT-1

Casio marketed its small electronic keyboards as
"personal musical instruments". Casio keyboards
gained fame when European pop group Trio
played one on their hit "Da da da". Despite its
small size (12.8 x 3.4 x 1 inches) the PT-1 offered
a good number of musical functions, including ten
different rhythms and the ability to record a
sequence of up to 100 notes. It came in a variety
of colours, but white and black appear to be the
most common. Nothing quite like the Casio PT-1
has been made since. Every musical home should
have one.

£10–£20 / US$20–$35

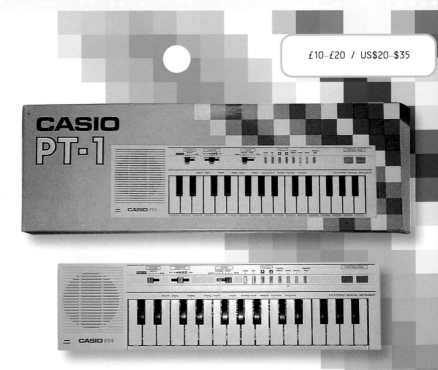

1987
EGA MASTER SYSTEM

ega launched its Master System after Nintendo's NES had established itself
 the dominant gaming console of the time. As a result, the Master System
as not successful in the USA, where Nintendo already owned 90 percent of
e market. Like every other console of the time, the
aster System used game cartridges, which were
serted in a slot at the top of the machine. The
nsole was unusual because it could accommodate a
cond game format, "Sega Cards", which fitted into
slot at the front of the machine. After its
parture, the Master System's architecture was
born in Sega's portable gaming machine, the
me Gear.

£15–£35 / US$30–$65

1988
SINCLAIR SPECTRUM +3

£30–£60 / $55–$110

In 1986, Sinclair's computer business was taken over by Amstrad, who was already making
its own computers. Amstrad released a version of the Spectrum before the +3: it was, as
you might guess, called the +2. However, the +3 was the last computer made under the
Spectrum name. Amstrad was not ashamed to market its Spectrums as pure gaming
machines and produced boxed sets with tie-ins
– James Bond movies, for example. The +3
used Amstrad's three-inch floppy disk drive
for storage, but despite the success of its
predecessor, it didn't sell well. This may
have been because of its high price: at the
time, just a little more money would have
bought you a more advanced machine.

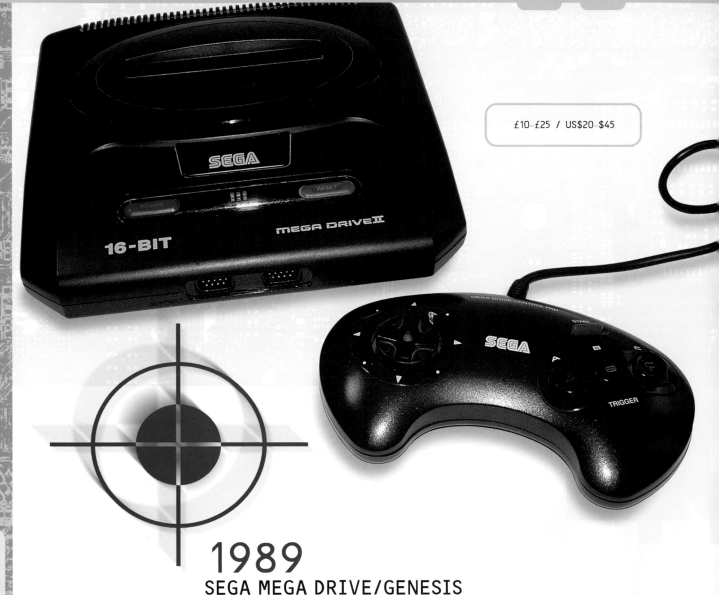

£10–£25 / US$20–$45

16-BIT

SEGA

MEGA DRIVE II

1989
SEGA MEGA DRIVE/GENESIS

After losing out to Nintendo's NES, Sega's next attempt at the console market was a winner. The Genesis (also known as the Mega Drive) was the first 16-bit game console launched in the USA at just under $190 (£105). Sega's new console came in a stealthy black colour, which was in stark contrast to Nintendo's grey NES. The Genesis was given a boost with the arrival of games made under licence from Disney. The console in the picture is the Mark II model that, like the original version, could be attached to a CD drive. A lack of third-party support and the arrival of new consoles meant that not many CD games were produced.

Which handheld electronic game was the most important of the twentieth century? Probably Nintendo's revolutionary Game Boy. This small machine, with its LCD screen and cartridge-based system, set the standard for mobile gaming. In fact, many would say that Nintendo has led the race all the way to the present day. Although it was capable of monochrome images only, the Game Boy's trump card was the range of available games; many were versions from Game Boy's big brother, the NES. This portable gaming classic was made in its millions, so it's very easy to find one today. Still a perfectly playable machine, the Game Boy should have a place in every technology collection.

£100–£150 / US$185–$280

1989
SAM COUPE

The Sam Coupé is often regarded as a successor to the Sinclair Spectrum, even though Sinclair had nothing to do with its development. Most other new computers in 1989 were 16-bit machines, but the Coupé was a Z80A-based 8-bit product. The first versions had 256KB of RAM and used tapes for storage, although later versions had more memory and 3.5-inch floppy disk drives. The Coupé was never a big seller and suffered from a series of technical problems. Today, though, it has a good following and decent examples of the Coupé are very difficult to find, which makes it a good collector's item.

£10–£15 / US$20–$30

1989
ATARI LYNX

The Atari Lynx was launched in the same year as Nintendo's Game Boy and, on paper, it looked like a better machine. It had more power and a back-lit colour screen, compared to the Game Boy's reflective monochrome version. Unfortunately, it was also much bigger, had fewer available games and its back-lit screen shortened battery life. It was also much more expensive at $200 (£110) compared to $110 (£60) for the Game Boy. The Lynx is a beautiful item and nowhere near as easy to find as the original Game Boy.

£20–£50 / $35–$95

1992
BARCODE BATTLER

Barcode Battler was marketed in different countries by different makers and was a peculiar game. To play it, you had to swipe barcodes through a slot on the front of the console. Barcodes were supplied on cards with the game, but any barcode from any product could be swiped, provided it fitted through the reading slot. With barcodes appearing on everything by 1992, the potential for a craze was there, but the game wasn't a success – even though it was reportedly popular in Japan. Boxed examples are fairly easy to find, suggesting that they were never played with.

£10–£15 / US$20–$30

£20–£40 / US$35–$75

1992
SEGA GAME GEAR

When the Game Gear was launched, Sega hoped to succeed where Atari had failed. Sega's machine was a back-lit colour-screen handheld that used game cartridges. In essence, the Game Gear was a portable Sega Master System, which meant that the library of games for the Master were easy to port to the Game Gear console. The case was much more rounded than Atari's Lynx, but quite large compared to the tiny, market-leading Game Boy. With a 3.2-inch screen and good game support, the Game Gear had some winning qualities. However, its size and poor battery life meant that it still came second to the Game Boy. Game Gears are relatively easy to find but buyers should check that the sound works properly.

£10–£20 / US$20–$35

1994
PLAYMATES STAR TREK PHASER

Playmates released this model of a Starfleet Phaser in 1994 as part of a range of items from the famous science fiction franchise. The range included weapons, science instruments and anything else that lit up or made a noise. The Phaser had working lights and sounds, so it was a fantastic item to hold and play with. It is highly likely that as many ended up in adult hands as children's! "Star Trek" has an enormous following across the world and practically anything to do with the TV series and films is attractive to collectors.

1995
SEGA SATURN

The Sega Saturn was launched at a price of $400 (£215), ahead of schedule and in an apparent rush. Perhaps Sega knew that Sony would launch its first game console, the PlayStation, within the year. The Saturn was a complex machine, with an architecture that made games' development difficult. As a result, some games failed to tap the full potential of the machine. With limited supplies, a lack of game titles and the arrival of the Sony PlayStation, the Saturn was never going to be a success. Today, good examples of the console are fairly scarce.

£15–£40 / US$30–$75

£15–£40 / US$30–$75

1996
NINTENDO 64

The Nintendo 64 was the last game console to use cartridges and arrived at a time when other consoles were already using CDs. Nintendo's strong game library was always an advantage and the 64 was launched with some very good titles, such as Super Mario 64. It set the standard for 3-D gaming. The games were truly amazing to behold, with characters walking in a three-dimensional virtual world. It brought new gaming elements to an audience brought up on 2-D platform titles.

£5–£15 / US$10–$30

1996
TAMAGOTCHI

Crazes are interesting phenomena. From time to time, the whole world seems to go mad for one particular thing, usually something cheap and small. In 1996 it was the Tamagotchi (translated as "watch inside an egg"), launched by Japanese firm Bandai. As the craze swept the globe, other manufacturers started to copy the format, which created an interesting and varied environment for collectors. In essence, the Tamagotchi owner hatched a virtual pet, usually from an egg, and then looked after it until it expired, after which they could hatch a new one. The "pet" was fed and played with, but if neglected, fell sick and potentially died young.

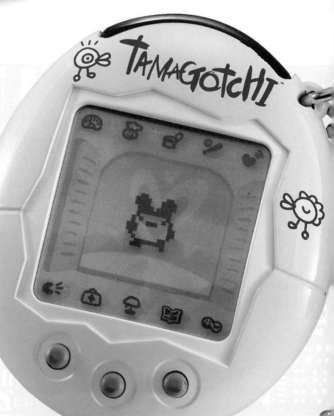

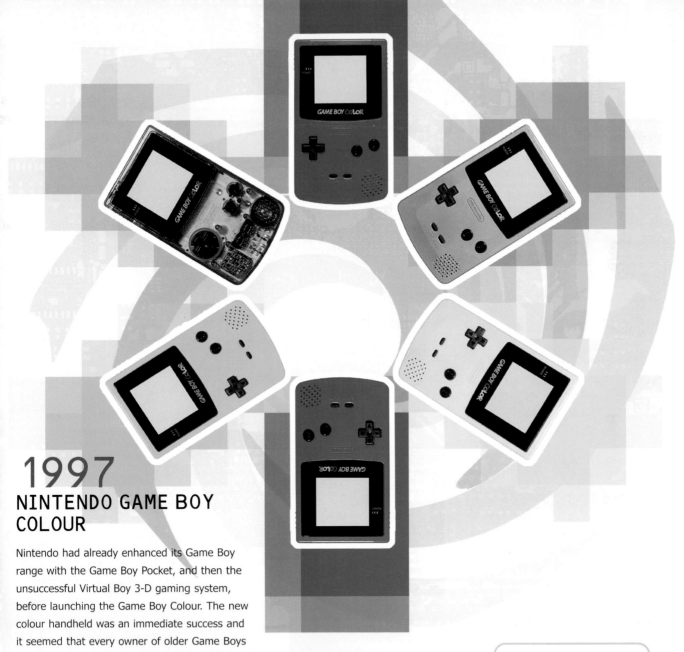

1997
NINTENDO GAME BOY COLOUR

Nintendo had already enhanced its Game Boy range with the Game Boy Pocket, and then the unsuccessful Virtual Boy 3-D gaming system, before launching the Game Boy Colour. The new colour handheld was an immediate success and it seemed that every owner of older Game Boys switched immediately. All previous Game Boy games worked on the new console, which made it all the more attractive, although only games written for the colour console offered full use of the new screen. The screen used a reflective colour LCD that could be tricky to view clearly. It was released in various colours, including some intended to attract girls.

£15–£25 / US$30–$45

1998
SEGA DREAMCAST

Sega launched the Dreamcast as the successor to its Saturn console. It was the first 128-bit console and promised astonishing graphics, with three-dimensional play. At under $200 (£110), the Dreamcast was a very good machine with a good range of games. However, Sega's timing was poor. Rumours of a new Sony machine, the PS2, and memories of the ill-fated Saturn led to the eventual demise of the Dreamcast. With it went Sega's involvement in gaming hardware: from then on, Sega concentrated solely on software and games. A decent number of Dreamcasts were sold and they are well worth collecting.

£5–£15 / US$10–$30

1998
TIGER FURBY

Tiger, an established toy manufacturer, released the electronic pet Furby on an unsuspecting world in the late 1990s. This strange hairy beast was touted as having artificial intelligence and, if you've ever seen one in action, you could have believed it. Tiger may have developed the toy in the hope of capitalizing on the success of Tamagotchi pets. Furbys came in many different colours. They would squawk and rock, demand food, sleep and, when placed in the vicinity of another Furby, talk to each other in a strange warble. Furby was weird but it sold by the million. Owners tended to look after their electronic pets and, as a result, many are sold on today complete with their original box and instructions.

£20–£40 / US$35–$75

2000
SONY AIBO

Aibo, which means "companion", was the name Sony gave to its entertainment robot, which it built in the form of a dog. Sony is one of many companies spending vast sums on robot research, and Aibo is just one of the outcomes. The Aibo shown here is the ERS-210, one of the first models. The robot was a very expensive item, not least because of the incredible amount of technology within its body. Aibos were capable of walking, dancing, seeing and hearing. Later models were able to pick up specially made bones. Aibos are truly unique and special pieces of technology that command high prices.

£400–£1,000 / US$740–$1,855

2001
NINTENDO GAME BOY ADVANCE

Nintendo repeated its 1989 achievement again in 2001: it launched the best handheld game yet made. The Game Boy Advance was a quantum jump from the previous model, in both design and technology. More buttons, a bigger screen and a new format for the game cartridges were just part of the overhaul. The design didn't make the old games obsolete since the old-style of cartridge worked on the new machine, even if they did stick out a little. At its heart was a 32-bit processor capable of driving a new generation of multi-dimensional games. The screen was a disappointment: there was no backlight and it had to be viewed at a specific angle for the best performance. Some owners modified their consoles and added their own back-lit systems. It was the best handheld game ever made – until it was overshadowed in 2003.

£30–£40 / US$55–$75

£70–£100 / US$130–$185

2002
SONY PS2

Sony joined the games console business late in the day.
However, its first console, the original PlayStation, was a huge
success and was always going to be a hard act to follow. To
maintain its market lead, Sony needed to create something
very special for its successor – and that is exactly what it did.
The PS2 was four times more powerful than its predecessor
and had a very attractive new feature: the ability to play DVDs.
It also played the original PlayStation games, which made the
new console instantly attractive to owners of the existing
machine. The PS2 offered the future prospect of playing
networked games over the Internet. On top of all of this, the
machine's design was a work of art.

2002
NINTENDO GAMECUBE

Nintendo was still relying on a cartridge-based console, the Nintendo 64, as the world moved into the new millennium. Meanwhile, Nintendo's competitors had CD machines in the shops, threatening the Japanese company's market share. So Nintendo's new machine, the GameCube, did use a disc format, but it was a proprietary smaller disc format. This allowed the console to be a very compact cube. A clip-on LCD screen could be purchased to turn the GameCube into a practical portable gaming machine. Top-quality game titles were used to promote the new console, making it an immediate success – no mean feat given the intense competition at the time. It looked like a two-horse race between Sony and Nintendo. Then Microsoft arrived on the scene...

£35–£50 / US$65–$95

2002
MICROSOFT XBOX

£50–£80 / US$95–$150

Microsoft had never been associated with gaming machines before, but that didn't stop the company creating a classic console at its first attempt. Launched at less than $300 (£160), the XBox offered a new level of graphics performance and games that made the most of it. The console race was suddenly one with three very fast horses. The price of new consoles tended to drop soon after release and the XBox was no different. Microsoft's first games console has guaranteed status as a future classic to be sought out by collectors.

2003
PRODUCT ENTERPRISE TALKING DALEK

Daleks were the arch-enemies of BBC TV's time-travelling Doctor Who. These hideous creatures needed to live and travel in machines, the robot-like monster in the photo. Designed in 1963, Daleks were constructed of cheap materials, hence the sink-plunger arm and car indicator lamps on the head. Daleks have starred in many "Doctor Who" television episodes and two movies. Early Dalek toys from the 1960s are highly sought-after items. This rather good model by Product Enterprise is a good bet for collectors, not least because of the scheduled reappearance of Doctor Who and the Daleks on television.

2003
NINTENDO GAME BOY SP

Nintendo introduced the Game Boy SP relatively soon after its Advance version. The SP is a better handheld than the Advance in most respects. The poor reflective screen of the previous model was now a marvellous back-lit screen that folded in a clamshell-style case. The new SP had an internal rechargeable battery and could play all previous versions of Game Boy cartridges. It was released in a variety of colours and one was even painted in the style of Nintendo's 1985 NES console. Measuring just 3.1 x 3.1 x 0.9 inches when folded, the SP was very compact. Quite simply the best handheld electronic game made to date, it is a must-have for collectors.

£40–£60 / US$75–$110

RESOURCES

Web Sites

apple2history.org
Apple computers history

darkwatcher.psxfanatics.com
Game console history

inventors.about.com
Inventor history & resources

oldcomputers.net
Obsolete technology

www.8trackheaven.com
8-track history

www.amstrad.dk
Amstrad history & resources

www.atariage.com
Atari history

www.atarimuseum.com
Atari history

www.bang-olufsen.com
Audio & television equipment

www.beoworld.co.uk
Bang & Olufsen history

www.binarydinosaurs.co.uk
Home computing history

www.brionvega.it
Audio & television equipment

www.bt.com
Telecommunications history

www.classicgaming.com
Video game history

www.commodore.ca
Commodore history

www.consoledatabase.com
Game console history

www.datamath.org
Historical calculators

www.designaddict.com
*Trading site for design
collector's items*

www.ebay.com
Online auction site

www.ericofon.com
Ericofon history

www.gameandwatch.com
*Nintendo Game & Watch
information*

www.heathkit-museum.com
Heathkit history

www.hpmuseum.org
Hewlett Packard history

www.iainsinclair.com
Designer, Eon Torches

www.intellivisionlives.com
Mattel Intellivision history

www.ipdb.org
Internet Pinball Database

www.jacob-jensen.com
Designer

www.jeffbots.com
Toy robots

www.labguysworld.com
*Early video recording
equipment*

www.laserdiscarchive.co.uk
Laser disc history

www.ledwatches.net
Historical electronic watches

www.linn.co.uk
Audio equipment

www.mathmos.com
*Lava lamps & other lighting
products*

www.minidisc.org
Minidisc information

www.naim-audio.com
Audio equipment

www.nvg.ntnu.no/sinclair
Planet Sinclair

www.oldcalculatormuseum.com
Historical calculators

www.old-computers.com –
Historical computers

www.pinballmachine.pwp.blue
yonder.co.uk
Pinball machine information

www.pong-story.com
Video game history

www.pure-digital.com
DAB radio manufacturer

www.radiomuseum.org
Radio history

www.recording-history.org
*Sound recording technology
history*

www.retrocom.com
CB radio history

www.retrogames.co.uk
Handheld game history

www.retrokit.co.uk
Vintage collector's items

www.sam-coupe.co.uk
*Sam Coupé History &
resources*

www.seiko.co.jp
Seiko history

www.synthmuseum.com
*Electronic musical instrument
history*

www.system16.com
Arcade game history

www.theledwatch.com
Electronic & digital watches

www.tvhistory.tv
Television history

www.vidipax.com
*Early video recording
equipment*

www.vintagecalculators.com
Historical calculators

www.vintageradio.co.uk
Vintage radio experts

www.vintage-technology.info
Vintage technology reference

www.voidware.com
Historical calculators

Books

Classics of Design, Hilary
Beyer & Catherine McDermott,
Brown Reference Group, 2002

*Complete Collector's Guide to
Pocket Calculators, The*, Guy
Ball & Bruce Flamm, Wilson
Barnett Publishers, 1997

Look of the Century,The,
Michael Tambini, Dorling
Kindersley, London 1996

Sinclair Story, The, Rodney
Dale, Duckworth, London
1985

Sound Design, David
Attwood, Mitchell Beazley,
London 2002

INDEX

Altair 8800 computer 22
Ampex VR-660B video
  recorder 65
amplifiers 74, 83
Amstrad: CPC-464 computer
  139
  PCW 8256 word processor
  46
Apple: Apple II computer 29
  iBook G4 laptop 58
  iMac computer 51
  iPod music player 105
  Lisa computer 40
  Macintosh Classic 43
  Newton PDA 49
Atari: 400 home computer
  123
  2600 video console 118
  Lynx game machine 146
  Portfolio pocket PC 47

Bandai games: Galaxian 129
  Missile Invader 126
Bang & Olufsen: Beogram
  4000 78
  Beomaster 1900 85
Barcode Battler 147
Baygen Freeplay wind-up
  radio 97
Blackberry mail service 55
Blueroom Minipod speakers
  101
Braun: ET-23 calculator 22
  Voice Control Alarm 92
Brionvega: Algol TV 66
  RR126 music centre 68
  TS502 radio 66
Bulova Accutron watch 64

Calcu-Pen 23
Cambridge Computers Z88
  47
cameras 79, 103–4, 106
Camputers Lynx computer
  132
Canon: Pocketronic 15
  X07 portable computer
  41
Casio: CQ-1 calculator/alarm
  26
  Exilim Z3 camera 106
  FX-120 calculator 23
  FX-2000 calculator 29
  Mini calculator 17
  Missile Defender game
  135
  MX-10 home computer
  139

PB-700 pocket computer
  39
  PT-1 keyboard 142
  TV-400 LCD pocket TV 94
CB radio 88
CGL M5 home computer 130
Chess, electronic 114
Citizen : LCD pocket TV 93
  LCD watch 89
Coleco Pac-Man 129
Commodore: 64 computer
  131
  PET computer 30
  S61 Statistician 32

DAB radios 102, 104
Dalek toys 156
Digital PDP-6 minicomputer
  14
Dragon 32 home computer
  132

Entex Defender game 127
Eon Classic LED torch 100
Epson HX-20 38
Ericofon 69

Football, Electronics 119
Friden EC-130 calculator 14
Furby electronic pets 152

Gakken games 135, 136
Girard Perregaux Casquette
  LED watch 86
Goldstar MD-R1 Minidisc 95
Grandstand: Astro Wars 130
Firefox F-7 138
Grundig radios 78, 81

Handspring Treo Smartphone
  59
Heathkit GR-98 Airband
  Receiver 77
Hewlett-Packard: HP-01
  calculator/watch 31
  HP16C calculator 37
  HP-35 calculator 17
  HP-65 calculator 21
  HP-71B calculator 42
  HP 110 portable
  computer 44
  HP 95LX palmtop 48
  IPAQ pocket PC 56
IBM: Personal Computer 36
  portable computer 42
Intertec Superbrain 36
iPod digital music player 105

Jupiter Ace computer 137
JVC Videosphere TV 75

K-9, talking toy 115
Karlsson Nixie tube clock 100
keyboard, electronic 142

Lava Lamp 67
Leica Digilux 1 camera 103
Lexon Tykho Radio 98
LG Phenom palmtop 52
Linn Sondek LP12 turntable
  80
Logic 5 game 115

Magnavox: 8-track receiver
  77
  Odyssey video console
  113
Master Mind, electronic 119
Masudaya Robby the Robot
  140
Mattel: Electronics Football
  119
  Intellivision console 126
Microsoft XBox console 155
Milton Bradley: electronic
  games 115, 121
  Vectrex games console
  133
Minidisk player/recorder 95
Motorola 8500X cell phone
  44
Musical Fidelity X-Ray CD
  Player 97

Naim NAP250 amplifier 83
Nintendo: 64 console 150
  Entertainment System 141
  Game & Watches 127,
  134
  Game Boys 145, 151,
  153, 157
  Gamecube 155
Nixie tube clock 100
Nixon The Dork watch 107

Olivetti Divasumma calculator
  19
Olympus DM-1 Voice
  Recorder 53
Omega Time Computer
  watch 84
Oric Atmos home computer
  138
Osborne portable computer
  37

Pac-Man game 129
Palitoy: Merlin 123
  Talking K-9 115
Palm Pilots 50, 53
Panasonic Toot-a-Loop 112
Philips: CD100 CD player 90
  GF303 record player 75
  N1502 video recorder 82
  VLP700 LaserVision 89
pinball machine 125
Playmates Star Trek Phaser
  148
Polaroid SX70 Land Camera
  79
President CB radio 88
Prinztronic Mini 24
Psion: 3 pocket computer 48
  5 pocket computer 48
  Organiser 43
  Wavefinder DAB radio 102
Pulsar: calculator watch 25
  Time Computer LED
  watch 80
Pure Digital Evoke-1 DAB
  radio 104

Radio Shack TRS80 1 35
Ragen Microelectronic 16
Research Machines RM-380Z
  39

Sam Coupé computer 145
Science of Cambridge MK14
  117
Sega: Dreamcast 152
  Game Gear 147
  Master System 143
  Mega Drive/Genesis 144
  Saturn 149
Seiko watches: Data 2000
  41
  LCD 86
  speaking 95
Sharp: EL-8 calculator 16
  EL-8130 calculator 31
  MZ80K computer 34
  PC-1211 pocket computer
  34
  Zaurus C750 PDA 54
Simon game 121
Sinclair: Black Watch 85
  C5 vehicle 93
  Cambridge calculator 20
  Enterprise Programmable
  33
  Executive calculator 18
  Flat Screen TV 92
  Micro FM radio 68

Micromatic radio 71
Microvision pocket TV 87
Neoteric 60 amplifier 72
Oxford calculator 25
PDM35 multimeter 33
President calculator 27
QL computer 45
Series 2000 amplifier 73
Sovereign calculator 28
Spectrum +3 computer
  143
Stereo Sixty 74
Wrist Calculator 32
ZX80/ZX81 computer 128
ZX Spectrum computer
  131
Sony: Aibo robot dog 153
  Clié PEG-NR70V pocket
  PC 57
  DSC U20 digital camera
  104
  HB-101 home computer
  141
  ICR-100 radio 72
  MZ-1 Minidisc player 95
  PS2 games console 154
  TC-124CS cassette player
  73
  TR650 radio 65
  Vaio-U3 Notebook 54
  Walkman 87
  Watchman pocket TV 91
Space Invaders 124
speakers, Minipod 101
Stax SR-X Ear Speakers 76
Stylophone 112
Swatch: Cellular phone 96
  Cordless phone 99
  The Beep 49
  watch 91

Tamagotchi 150
Tele-Tennis games 116
Texas Instruments games
  122
Tiger Furby 152
Tomy Blip game 120
Tomytronic 3-D game 136
torch, LED 100
Toshiba HX-10 computer 142
Trimphone 70
TSI Speech Plus calculator
  26
turntable 80

video recorders 65, 82

wind-up radio 97

CREDITS

The publishers would like to thank the following sources for their kind permission to reproduce the pictures in the book.

(T = Top, B= Bottom and M= Middle)

Apple Computer, Inc.: 13r, 58, 105; Atari Historical Society: 47b, 146; Terry Bailey: 77t, 88; Bandai: 150b; Bang & Olufsen: 78b, 85b; Tony Barnett: 145t; Steve Behan: 156; Braun: 22b; Brionvega: 66, 68b; Tony Brown: 81; BT Group Archives: 70; Alan Buck: 140; Stuart Campbell: 125; Casio: 94; Compumuseum.de: 48t; Computer History Museum: 12br, 14bl, 14bm, 17b, 37t; Corbis: /Letourneur Herve: 146, /Roger Ressmeyer: 40; Roger Diehl: 65b; Tiffany Dodd: 7; Ericsson Archive: 69; Rick Furr: 31t, 37b; Getty Images: /Malcolm Clarke: 25, /Urbano Delvalle/Time Life Pictures: 152b, /Microsoft: 155b, Joe Raedle/Newsmakers: 150; Grand Canyon: 77b; Hewlett Packard: 13bm, 44t, 56; HomeComputer.de: 130b, 139; IBM United Kingdom Limited: 36b; Fred Jan Kraan: 38; JVC: 75b; Ed Kearn / www.gateman.com: 36t; Erik Klooster: 133, 143t; Chris Klupak: 142b; Ledwatch.com: 84, 86b; Leica: 103; Lexon: 98; Linn: 80b; Dennis Marsden: 87; Steve Marshall: 138b; Simon Maurice: 135b; Mathmos: 67; Rik Morgan/www.handheldmuseum.com: 119, 120, 126t, 127t, 129, 130, 135t, 136t, 138t; Musical Fidelity: 97b; Peter Muckermann: 17t; Nintendo/Cake Media: 141b, 145, 151, 153b, 155t, 157; Nixon Europe: 107; Old-Computers.com: 39b, 41t; Olympus: 52b; Osmotec Living Concepts: 100t; Palm One: 53b; Philips International B.V./Philips Company Archives: 6, 75, 89b, 90, 91r; Picture-Desk/The Advertising Archive: 51; Psion: 102; Pure Digital: 105b; Research: 39t; Rex Features: 93t, 149, 152, /Scott Aiken: 55, /Peter Brooker: 124; Rundfunkmuseum.fuerth.de: 78t; Scandyna: 101; Science Photo Library: Andrew Syred: 4t, 11, /TEK Image: 4m, 61, /Victor Habbick Visions: 4b, 109; Science & Society/Science Museum: 8l, 22t, 29t, 30, 35, 45, 49b, 50b, 64, 71, 79, 82, 87t, 91l, 92t, 97t, 118, 132t, 137, 141t; Seiko UK Ltd: 41b, 86t, 95; Sharp Corporation: 31b, 34, 54b; Wolfgang Shefer: 114; Iain Sinclair:100b; Steven Smith: 115t; Tom Spilliart: 115b, 121, 123b, 126b, 136b; Sony: 54t, 57, 72b, 104t, 153t, 155; Stax: 76; Malcom Surl: 46; Swatch: 49t, 99; Enrico Tedeschi/Electronic Icons: 47t, 65t, 68t, 117, 123t, 128, 131, 132b, 142t, 143b; Texas Instruments: 122; Viktor T. Toth: 42t; Pepe Tozzo: 8r, 14t, 18, 20, 21, 25b, 27, 29b, 33, 44b, 49b, 50t, 52, 59, 72t, 73, 74, 80t, 85t, 89t, 91r, 92b, 93b, 96, 106, 116, 114; Lawrence Tunney: 83; Harry Viannakis: 126b, 134; Vintage Calculators/Nigel Tout: 12bm, 15, 16, 23, 24, 26,, 28, 32, 33t;

David White: 148; Tony & Sarah Whitehead: 147t; www.TcoCD.de:19; www.tonh.net/museum: 43b

Special thanks to Roger W. Amidon, Tony Barnett, Dan Bellander, Rick Bensene, Sylvain Bizoirre, Wes Brummett, Daniel Eckes, Pablo Gargano, David Gesswein, Anders Gidlöf, Adrian Graham, Frank Guenthoer, Ton den Hartog, Fred Jan Kraan, Wayne Kneeskern, Erik Klooster, Oliver W. Leibenguth, Hiromi Morita, Gino Mancini, Dennis Marsden, Karl Morris, Reza Ramdjan, Anel Rodriguez, Carolyn Potts, Mizue Sakurai, Stan Sieler, Tom Spilliaert, Malcolm Surl, Nigel Tout, Viktor T. Toth, Pepe Tozzo, Stefan Walgenbach, Gerd Walther, David White and Harry Yiannakis.

Every effort has been made to acknowledge correctly and contact the source and/or copyright holder of each picture, and Carlton Books Limited apologises for any unintentional errors or omissions, which will be corrected in future editions of this book.

NOTE FROM THE AUTHOR:

There are scores of people to whom I owe thanks, too many to list here, but there are some to whom I owe special gratitude. These are Andrea and Christian Tozzo, who I should thank for their endless support and patience, Katherine Higgins, who got me involved in this project in the first place and Terry Bailey, who policed my poor grammar.